I0043548

William Curtis

Places and descriptions of such plants as grow wild in the environs of London

William Curtis

Places and descriptions of such plants as grow wild in the environs of London

ISBN/EAN: 9783742858047

Manufactured in Europe, USA, Canada, Australia, Japa

Cover: Foto ©berggeist007 / pixelio.de

Manufactured and distributed by brebook publishing software
(www.brebook.com)

William Curtis

Places and descriptions of such plants as grow wild in the environs of London

FLORA LONDINENSIS:

OR

PLATES and DESCRIPTIONS of such PLANTS

AS GROW WILD IN THE

ENVIRONS of LONDON:

WITH

Their Places of *Growth*, and Times of *Flowering*; their feveral
Names according to LINNÆUS and other Authors:

WITH

A particular DESCRIPTION of each PLANT in LATIN and ENGLISH.

To which are Added,

Their feveral Ufes in *Medicine*, *Agriculture*, *Rural Oeconomy*, and other *Arts*.

By W I L L I A M   C U R T I S,

DEMONSTRATOR of BOTANY to the COMPANY of APOTHECARIES.

VOL. I.

LONDON:

Printed for and Sold by the AUTHOR, No. 51, *Gracechurch-Street*; and B. WHITE, *Bookfeller*, in *Fleet-Street*.

MDCCLXXVII.

THE

PREFACE.

ALTHOUGH the Author does not here mean to give a Preface at large, referving that until the firft volume, containing thirty-fix numbers or two hundred and fixteen plants, fhall be completed; yet he prefumes it will be fatisfactory to his fubfcribers, and the public, to be informed a little more fully of the nature and defign of the work; as it will alfo give him an opportunity of anfwering fome few objections that have been made to the plan of it.

The primary defign of it then, is to facilitate a knowledge of the plants of our own country, and eftablifh each fpecies and variety on a firm bafis: this the Author confiders as the grand defideratum at prefent; this arduous tafk once accomplifhed, a way will be opened, and a foundation laid for numberlefs improvements in Medicine, Agriculture, &c.

To be enabled to do this, he means to take the greateft pains in the examination of thofe plants which he figures; to have them drawn from living fpecimens moft expreffive of the general habit or appearance of the plant as it grows wild; to place each plant, as much as is confiftent, in the moft pleafing point of view; and to be very particular in the delineation and defcription of the feveral parts of the flower and fruit, more efpecially where they characterize the plant.

And in order that he may obtain a more perfect knowledge of each plant; that he may fee it in every ftage of its growth, from the germination to the maturity of its feed; that he may compare and contraft the feveral fpecies together; that he may make experiments to elucidate the nature of fuch as are obfcure, or bring into more general ufe thofe which bid fair to be of advantage to the public; he is now cultivating each of them in a garden near the city, into which, by the kind affiftance of his friends, he has already introduced, in the courfe of one year, about five hundred different fpecies, including fixty of that moft valuable tribe of plants the graffes.

Although the afcertaining and fixing of the plants will be his principle object, yet to make the work more ufeful to the public, as well as inftructive and entertaining to the young botanift, his utmoft endeavours will be ufed to lay before them whatever may be found *ufeful* in old botanic writers; and here they muft not be furprized to find many of the numerous and imaginary virtues, which they attributed to almoft every plant, purpofely omitted: the difcoveries made by modern authors, particularly relative to *Agriculture* and *Rural Oeconomy*, will be carefully attended to; as here feems to be a field juft opening to view, from whence the public is likely to draw great and lafting advantages: and as a knowledge of the plants themfelves is firft neceffary, and for want of which, indeed, the experimental farmer cannot effectually communicate his improvements, he finds himfelf peculiarly happy in contributing his fhare to the public good.

He is neverthelefs fenfible how inadequate his abilities, or indeed the abilities of any *one* perfon are, to render a work of this kind any ways complcat; he therefore refpectfully folicits the affiftance of thofe, who wifh well to the improvement of *Englifh Botany* and *Englifh Agriculture:* any information they fhall be pleafed to communicate, fhall with thofe favours he has already received from divers of his friends, be gratefully acknowledged; and to induce them the more readily to communicate, he has fubjoined a catalogue of thofe plants which (with many others) are already drawn, and which he intends fhall form the next Fafciculus.

He is forry it has not been in his power to publifh his numbers fo faft as was originally propofed: the delay has chiefly been occafioned by the lofs of one of his principal artifts, whofe place is now fupplied by two others equally eminent; fo that the drawing and engraving, which before fell to the fhare of one perfon, being now divided betwixt two, he flatters himfelf he fhall be able to publifh a number once a month, or fix weeks at fartheft—he is however determined never to facrifice the accuracy or utility of the work to hurry—on this principle he has been at the expence of having fome of his plates engraven twice, and even three times over before he could venture to publifh them. As the delay has originated from this fource, he hopes none of his fubfcribers that have hitherto fo generoufly contributed to the carrying on of the work, will withdraw that affiftance, which alone can enable him to profecute it with advantage to the public, credit to himfelf, and fatisfaction to them.

It now remains to obviate fome few objections which have been made to the plan of this work; and firft, it has been fuggefted to the Author, that it would have been better received, if, inftead of purfuing the prefent plan, he had publifhed thofe plants only which were not figured in the *Flora Danica,* a work now carrying on in *Denmark* under the aufpices of the *King*: but a few moments reflection, muft he prefumes be fufficient, to convince every unprejudiced perfon, how inadequate fuch a partial publication would have been to the making a knowledge of the plants of our country more general among ourfelves—at beft fuch a work could only anfwer the purpofe of thofe few individuals who are in poffeffion of that part of the *Flora Danica* already publifhed; and as that is ftill going on, there is no doubt but the fame plants would be publifhed by both Authors; thus, the *Butomus umbellatus, Solanum Dulcamara,* and *Ervum hirfutum,* have been publifhed in the *Flora Danica* fince they were publifhed in the *Flora Londinenfis,* fo that in the end even thofe perfons would be obliged to purchafe duplicates of the fame plant.

Another reafon why the Author could not adopt the plan propofed to him, was the limited fcale of the *Flora Danica,* which contains the figures and names of the plants only, but gives us no account of their properties, nor teaches us how to diftinguifh the difficult plants from one another; the plates likewife being *fmall folio,* cannot admit many of the plants of their natural fize, feveral of the graffes for inftance, as the *Feftuca fluitans* and *Aira aquatica* are obliged to be fo cut and diminifhed as fcarcely to be known. Many other objections might be urged without any view to depreciate a work which, though not fo complcat in fome refpects as could be wifhed, has exceeding great merit:—but thefe will probably be deemed fufficient.

The

The PREFACE.

The engraving of one plant only on each plate has been another objection which some have strongly urged, while others have in as warm terms testified their approbation of it. It may be proper to mention, that whether one or more had been engraven on a plate, the difference in the expence would have been trifling, and chiefly in the paper: as they now are, each is distinct, and every one is at liberty to place them according to that system which he most approves of.

The want of figures of reference to the plates, or letter-press, has been perhaps a more solid objection; but the Author hopes, that by the use of the indexes described below, this also will be obviated.

Having now, so far as he can recollect, answered every thing deserving the name of an objection, he willingly submits his performance to the judgment of a candid and impartial public; conscious of having used his best endeavours to be serviceable in his department.

Uses of the *Indexes*, with *Directions* for *Binding*.

In the first Index the plants are placed according to the System of LINNÆUS, with which it is presumed, the greatest part of his subscribers are best acquainted. To find out any plant, even though the person be not acquainted with this mode of arrangement, look in the alphabetical English or Latin Index, and you will find the figures corresponding with them as placed in the book: if for example I want to find Ivy, I look for it in Index, No. 3, where the English names are alphabetically arranged, and find it to be the 16 plate, as there are 72 plates in each Fasciculus, I can readily guess within a few plates where it is placed: to those who have been accustomed to look out plants in LINNÆUS's works it will come easier; but if each subscriber will take the small pains of figuring the plates with a black lead pencil, any plant may then be immediately referred to. The Author could not hit on any mode more eligible, consistent with the irregular order in which he has been obliged to publish his plants.

With every third Fasciculus will be given a general and more copious Index, with a Glossary of the technical terms used in the work.

He would recommend to his subscribers, that each Fasciculus containing twelve numbers, be *bound in boards, and not cut at the edges*; the plates to be placed in the same order in which they occur in the first Index; taking care that each plate be put opposite to the letter-press belonging to it, with a leaf of thin paper betwixt them. If any should be at a loss to have them properly done, they will be pleased to send them to *Rabam Reepe's*, Bookbinder, in Crooked Lane, near the Monument, who binds the Author's.

N. B. It may be necessary to caution the Bookbinder against beating the Numbers, as that operation would probably destroy the beauty of the plates.

A

Catalogue of those Plants which are *intended* to be Published in the next *Fasciculus*.

Anemone nemorosa	Geranium Columbinum	Plantago major
Adoxa moschatellina	Hyacinthus non scriptus	Plantago Coronopus
Ajuga reptans	Hyoscyamus niger	Plantago media
Aira præcox	Hypericum montanum	Poa rigida
Arabis thaliana	Hypericum quadrangulum	Poa compressa
Arenaria tenuifolia	Hypericum hirsutum	Polygonum amphibium
Achillea Ptarmica	Ilex Aquifolium	Polytrichum commune
Briza media	Iris Pseudacorus	Ranunculus hirsutus
Corylus avellana	Lamium amplexicaule	Ranunculus Ficaria
Chærophyllum sylvestre	Lysimachia nemorum	Sagina erecta
Convolvulus arvensis	Lysimachia nummularia	Saxifraga tridactylites
Circæa lutetiana	Lysimachia tenella	Spergula nodosa
Chenopodium Vulvaria	Lysimachia vulgaris	Sedum dasyphyllum
Dipsacus sylvestris	Ligustrum vulgare	Sedum reflexum
Epilobium angustifolium	Lotus corniculata	Symphytum officinale
Epilobium ramosum	Myosurus minimus	Sparganium erectum
Erica cinerea	Malva officinalis	Tussilago farfara
Fumaria officinalis	Malva minor	Tormentilla erecta
Festuca duriuscula	Medicago lupulina	Thymus serpyllum
Festuca myuros	Osmunda spicant	Trifolium fragiferum
Glechoma hederacea	Oxalis acetosella	Valeriana dioica
Geranium molle	Orchis Morio	Veronica officinalis
Geranium rotundifolium	Ornithopus perpusillus	Veronica hederifolia
Geranium perenne	Plantago lanceolata	Veronica arvensis

7.

INDEX I.

In which the Plants contained in the firſt Faſciculus are arranged according to the Syſtem of LINNÆUS.

I N D E X II.

In which the Latin Names of the Plants
are arranged Alphabetically.

I N D E X III.

In which the Englifh Names of the
Plants are arranged Alphabetically.

VERONICA AGRESTIS. PROCUMBENT GARDEN-SPEEDWELL.

VERONICA *Linnæi. Gen. Pl.* DIANDRIA MONOGYNIA.

Raii. Syn. Gen. 18. HERBÆ FRUCTU SICCO SINGULARI FLORE MONOPETALO.

VERONICA *agrestis*, floribus solitariis, pedunculatis; foliis cordatis incisis, petiolatis; caule procumbente.

VERONICA *agrestis*, floribus solitariis, foliis cordatis incisis pedunculo brevioribus. *Linnæi Syst. Vegetab. p. 56.*

VERONICA floribus solitariis, foliis cordatis incisis petiolatis. *Hudson Fl. Angl. p. 6.*

VERONICA caule procumbente; foliis petiolatis, ovatis, crenatis. *Haller. Hist. V. 1. n. 594.*

VERONICA *agrestis. Scopoli. Fl. Carn. p. 21* DIAGN. Primiflora; foliis ovato-cordatis, crenatis, pedunculo brevioribus.

VERONICA floribus singularibus, in oblongis pediculis, Chamædryfolia. *Raii. Syn. p. 279.* Germander-Speedwell or Chickweed.

ALSINE foliis Trissaginis. *Gerard. emac. 616. Parkinson. 764.*

ALSINE Chamædryfolia flosculis pediculis oblongis insidentibus. *Bauhin. Pin. 250. Oeder. Fl. Dan. Icon. 449.*

RADIX annua, fibrosa.

CAULES plures, primum erecti, tandem procumbentes, semipedales, subvillosi, teretes.

FOLIA alterna, ovato-cordata, serrata, petiolis brevibus insidentia, subhirsuta.

FLORES pedunculati, pedunculi axillares, longitudine fere foliorum, post florescentiam reflexi.

CALYX: PERIANTHIUM quadripartitum, laciniis lanceolatis, hirsutis, subtortuosis, *fig.* 1.

COROLLA monopetala, subrotata, calyce brevior, lævissime fere tactu decidua; TUBUS brevissimus; LACINIÆ concavæ, subrotundæ, nunc penitus cœruleæ, nunc venis cœruleis striatæ, *fig.* 2.

STAMINA: FILAMENTA duo, alba, medio crassiora; ANTHERÆ cœrulescentes, *fig.* 3.

PISTILLUM: GERMEN subcompressum, hirsutulum, basi nectario cinctum; STYLUS viridis, apice incrassatus, staminibus brevior; STIGMA album, capitatum, *fig.* 4.

PERICARPIUM CAPSULÆ Veronicæ serpyllifolia similis, at major rotundiorque, *fig.* 5.

SEMINA pallide fusca, plerumque 6 in singulo loculamento, rugosa, hinc convexa, inde concava, *fig.* 6.

ROOT annual and fibrous.

STALKS several, first upright, then procumbent, about six inches in length, round and somewhat villous.

LEAVES alternate, of an oval-heart shape, serrated, placed on short foot-stalks and slightly hairy.

FLOWERS placed on foot-stalks, which proceed from the Axillæ of the leaves and are nearly of the same length; after the flowers are gone off turning back.

CALYX: a PERIANTHIUM divided into four lacinia, which are lanceolate, hairy, and somewhat twisted, *fig.* 1.

COROLLA monopetalous, somewhat wheel-shaped and shorter than the Calyx, falling off on the least touch; the TUBE very short; the LACINIÆ concave, and roundish, sometimes wholly blue, sometimes striped with blue, *fig.* 2.

STAMINA: two FILAMENTS of a white colour and thickest in the middle; ANTHERÆ blueish, *fig.* 3.

PISTILLUM: GERMEN flattish, a little hairy and surrounded at bottom by a Nectarium; the STYLE green, thickest at top, and shorter than the Stamina; STIGMA roundish and white, *fig.* 4.

SEED-VESSEL a CAPSULE like that of the *Veronica serpyllifolia*, but larger and rounder, *fig.* 5.

SEEDS of a pale brown colour, generally 6 in each cavity, wrinkled, convex on one side and hollow on the other, *fig.* 6.

THERE are few Botanists but what are apt to confound this species of Veronica with the *Veronica arvensis*, and this appears to arise in some degree from their similarity to each other, but more perhaps from the similitude of their Latin, and the ambiguity of their English names. To prevent in some degree this confusion, I have taken the liberty of altering the English name of *Germander-Speedwell* or *Chickweed* to that of *procumbent garden Speedwell*, in order that the young Botanist may thereby more readily distinguish it from the species above mentioned. The stalks of the *Agrestis* are usually procumbent, and it is found generally in Gardens; whereas the *Arvensis* has an upright stalk, and with us is found most commonly on Walls. Besides such obviously distinguishing characters, these two plants differ confiderably in many other respects. In the *Arvensis* the leaves are sessile, in *this* they are placed on footstalks; in the *Arvensis* the flowers are sessile, in *this species* they likewise, are placed on foot-stalks: and a difference still more remarkable, or at least more curious, exists, which seems not to have been attended to, viz. the largeness and roundness of the seed-vessels, and the particular structure of the seed. In most of the Veronicas the seed-vessel is heart-shaped, and even in this species it retains somewhat of that form, although each of the Cavities is large and round; and if we examine the form of the seeds, we shall not wonder at this particular construction, for each seed instead of being small and flat as in other Veronicas, is large, convex on one side, hollow on the other, and wholly different in its appearance. This peculiarity of structure, shows what inconstancy there is in the parts of fructification, and how improper it would be to found a Genus on the particular form of any one of them, since those which are in general the most uniform, are sometimes subject to such uncommon variations. The number of seeds in each Capsule is generally about 12, LINNÆUS says 8, SCOPOLI from 16 to 20.

This species grows frequently in Gardens, and flowers through most of the summer months. No particular virtues or uses are attributed to it,

Veronica agrestis.

Veronica Chamædrys.

VERONICA CHAMÆDRYS. WILD GERMANDER.

VERONICA *Linnæi Gen. Pl.* DIANDRIA MONOGYNIA.

Raii Syn. Gen. 18. HERBÆ FRUCTU SICCO SINGULARI, FLORE MONOPETALO.

VERONICA *Chamædrys* racemis lateralibus, foliis ovatis rugofis dentatis feffilibus, caule bifariam pilofo. *Lin. Syft. Vegetab. p.* 57. *Fl. Suecic. p.* 6.

VERONICA foliis cordatis fubrotundis, hirfutis, nervofis, ex alis racemofa, *Haller. hift. n.* 536.

CHAMÆDRYS fpuria minor rotundifolia. *Bauhin. pin.* 249.

CHAMÆDRYS fpuria fylveftris. *Parkinfon,* 107.

CHAMÆDRYS fylveftris. *Gerard. emac.* 657. *Raii Syn.* 281. Wild Germander, *Hudfon. Fl. Angl. p.* 5. *Scopoli. Fl. Carniol. p.* 15. *(a)* OEder *Fl. Dan. icon.* 448.

RADIX perennis, repens, fibrofa.

CAULES numerofi, decumbentes, teretes, duri, *bifariam denfa hirfuti,* ramofi.

FOLIA cordato-ovata. oppofita, nunc feffilia nunc petiolis brevibus infidentia, ferrata, venofa, hirfutula,

FLORES numerofi, ad 20, cærulei, petiolati: *Petioli* BRACTÆA lanceolatâ fuffulti; RACEMI longi, nunc oppofiti nunc folitarii.

CALYX PERIANTHIUM quadripartitum, perfiftens, foliolis lanceolatis, hirfutulis, *fig.* 1.

COROLLA monopetala, rotata, tubus breviffimus interné ad inferiorem partem villofus, LIMBO quadripartito, plano, laciniis fubcordatis ad bafin venis faturatioribus ftriatis, inferiore anguftiore, *fig.* 2.

STAMINA: FILAMENTA duo apice incraffata, adfcendentia, *fig.* 3. ANTHERÆ fagittatæ, *fig.* 4. POLLEN album, *fig.* 6.

PISTILLUM: GERMEN compreffum glandula nectarifera cinctum: STYLUS declinatus, cærulefcens, STIGMA obtufum, purpureum, *fig.* 5.

PERICARPIUM: CAPSULA cordata, *fubcompreffa,* pallide fufca, *calyce paulo brevior,* ad marginem hirfutulum, *fig.* 7.

SEMINA: plura, compreffa, flavefcentia, *fig.* 8.

ROOT perennial, creeping, and fibrous.

STALKS numerous, fpreading, round, hard, *hairy on each fide,* hairs very thick together, branched.

LEAVES of an heart fhaped oval form, oppofite, generally feffile, fometimes ftanding on fhort footftalks, ferrated, veiny, and flightly hirfute.

FLOWERS numerous, to 20, of a bright blue colour, forming long RACEMI (which are fometimes oppofite, fometimes fingle), ftanding on *footftalks,* each of which is fupported by a longpointed BRACTÆA.

CALYX: a PERIANTHIUM divided into four fegments, and continuing, the fegments lanceolate and flightly hairy, *fig.* 1.

COROLLA monopetalous and wheel fhaped, the TUBE very fhort, internally villous on the lowermoft fide, the LIMB flat, and divided into four fegments, the fegments fomewhat heart-fhaped, ftriated at bottom with veins of a purple colour, the lowermoft fegment narrower than the reft, *fig.* 2.

STAMINA: two FILAMENTS, thickeft at top, rifing upward, *fig.* 3. the ANTHERÆ arrow-fhaped, *fig.* 4. the POLLEN white.

PISTILLUM: the GERMEN flattifh, furrounded at bottom by a nectariferous gland, *fig.* 6. the STYLE hanging downwards, blueifh; the STIGMA blunt, and purple, *fig.* 5.

SEED-VESSEL: a CAPSULE, heart-fhaped, *flattifh,* of a light brown colour, *a little fhorter than the calyx,* and flightly hairy at the edge, *fig.* 7.

SEEDS feveral, flat, of a yellowifh brown colour, *fig.* 8.

The flowers of this Veronica are the largeft and moft fpecious of all the Plants of that Genus which grow wild in this Kingdom; many plants with lefs beauty are cultivated in our Gardens with the greateft care.

The leaves have been recommended by fome writers as a fubftitute for *Tea.*

It bears a confiderable refemblance to the *Veronica montana,* but differs effentially from that plant in the fize of its Seed-veftels and the great number of flowers which it bears on its Racemi. See *Jacquin. Flor. Auftriac. Vol.* 2,

When growing wild the leaves are ufually feffile or placed on very fhort foot-ftalks, when cultivated they become larger and the foot-ftalks moderately long; a kind of monftrofity, which LINNÆUS has likewife obferved, is very frequent on the leaves at the extremity of the ftalk; which are collected into a very hairy white knob, on opening one of thefe I found two or three Infects in their Pupa or Chryfalis ftate, which moft probably would have produced fome fpecies of Fly. This appearance is very common at the latter end of Summer.

This is an early blowing plant, and grows very common on dry banks, under hedges, and in orchards; it flowers in May and June.

Veronica Serpyllifolia. little smooth Speedwell,

or Paul's Betony.

VERONICA *Linnei Gen. Pl.* Diandria Monogynia.

Raii Syn. Gen. 18; Herbæ fructu sicco singulari, flore monopetalo.

VERONICA *serpyllifolia* racemo terminali fubfpicato, foliis ovatis glabris, crenatis. *Linnei Syft. Vegetab. p.* 56. *Fl. Suecic. p.* 6.

VERONICA caule recto, foliis ovatis, glabris, crenatis, petiolis ex alis unifloris, breviffimis. *Haller hift. n.* 546.

VERONICA pratenfis ferpyllifolia. *Bauhin Pin.* 247.

VERONICA pratenfis minor. *Parkinfon.* 551.

VERONICA minor. *Gerard emac.* 627.

VERONICA foemina quibufdam, aliis Betonica Pauli Serpyllifolia. *I. Bauhin.* III. 285.

VERONICA *Raii Syn. p.* 279. *n.* 3. *Hudfon, Fl. Angl. p.* 4. *n.* 4 *Scopoli Fl. Carniol. V.*1. *p.* 12..*n.* 10 *OEder Fl. Dan. icon.* 492.

RADIX perennis, fibrofiffima.	ROOT perennial, and very fibrous.
CAULES numerofi, ad bafin repentes, dein erecti, fimplices, palmares, teretes, læves.	STALKS numerous, creeping at bottom, then growing upright, fimple, three or four inches high, round and fmooth.
FOLIA oppofita, fubcrenata, fubrotundo-ovata, rariter et obfolete ferrata, glabra, trinervia.	LEAVES oppofite, nearly uniting at bottom, *of a roundifb-oval form, here and there flightly ferrated, fmooth* and trinervous.
FLORES albi, venis cærulcis picti, fpicati, pedunculati, alterni, Bracteæ magnæ, ovatæ.	FLOWERS white, coloured with blue veins or ftripes, growing in fpikes on foot-ftalks alternately. Floral leaves large and oval.
CALYX: Perianthium quadripartitum, laciniis ovato-acutis, glabris, *fig.* 1.	CALYX: A Perianthium divided into four parts, the Segments of an oval pointed fhape, and fmooth, *fig.* 1.
COROLLA monopetala, rotata; tubus breviffimus; laciniæ fubcordatæ, inferiore anguftiore; fuperiore lacinia ftriis aut venis purpureis octo notata, lateralibus venis duabus, interiore penitus albo, *fig.* 2.	COROLLA monopetalous, wheel-fhaped, the tube very fhort, the fegments fomewhat heart-fhaped, the lower one narroweft; the upper fegment marked with eight purple veins or ftripes, the fide ones with two, and the lower one entirely white *fig.* 2.
STAMINA: Filamenta duo, alba, apice incraffeta, *fig.* 5, 6. Antheræ cærulefcentes.	STAMINA: two Filaments, white and thickifh towards the extremity; the Antheræ blueifh *fig.* 5, 6.
PISTILLUM: Germen fubcompreffum, Stylus albus, apice paululum incraffatus, perfiftens. Stigma capitatum, rubens, *fig.* 3.	PISTILLUM: the Germen flattifh, the Style white, a little thicker towards the extremity, and continuing. Stigma roundifh, and of a redifh colour, *fig.* 3.
NECTARIUM ad bafin germinis, ut in Veronica Chamædrys.	NECTARY at the bottom of the Germen as in the Veronica Chamædrys.
PERICARPIUM: Capsula fubcordata, fufca, pro magnitudine plantæ magna, *fig.* 4.	SEED-VESSEL: a Capsule fomewhat heart-fhaped, of a brown colour, and large in proportion to the plant, *fig.* 4.
SEMINA plurima, 60 numeravi, e luteo-fufca, fub-ovata, *fig.* 8.	SEEDS numerous, of a yellowifh brown colour, and fomewhat oval fhape, *fig.* 8. We counted 60 in one capfule.

No particular virtues are attributed to this little plant by Writers.

It is one of the leaft of the Veronicas, and occurs frequently in Meadows and Fields, and fometimes in Gardens, flowering in the Spring and Autumnal Months.

There is a great deal of delicacy in its bloffoms, but they are too minute to make its beauty confpicuous enough for the Garden.

Its fmall, round, fmooth and fhining leaves readily diftinguifh it from the other Speedwells.

Veronica serpyllifolia

Anthoxanthum odoratum. Sweet-scented

or Vernal Grass.

ANTHOXANTHUM *Linnæi Gen. Pl.* DIANDRIA DIGYNIA.
Calyx. Gluma bivalvis, uniflora. *Corolla.* Gluma bivalvis, acuminata. *Semen* unicum.

Raii Synop. Gen. 27. HERBÆ GRAMINIFOLIÆ FLORE IMPERFECTO CULMIFERÆ.
ANTHOXANTHUM *odoratum* fpica oblonga, ovata, laxa.
ANTHOXANTHUM *odoratum* fpica oblonga, ovata, flofculis fubpedunculatis arifta longioribus, *Linnæi Syft.*
Vegetab. p. 67. *Fl. Suecic. No.* 33.
AVENA diantha, folliculo vilofo, calycis glumis inæqualibus, altera de imo dorfo, altera de fummo
ariftata. *Haller. hift. helv. No.* 1291.
ANTHOXANTHUM *odoratum Scopoli Fl. Carniol. No.* 39. *Hudfon Fl. Angl. p.* 10. *Stillingfleet mifcel.*
t. 1. *Schreber Gram. tab.* 5. *p.* 49.
GRAMEN pratenfe fpica flavefcente. *Bauhin. Pin.* 3.
GRAMEN vernum fpica brevi laxa. *Raii Syn.* 389. *Scheuch. hift.* 88.

RADIX perennis, fibrofa.

CULMI primum obliqui, demum erecti, dodrantales aut
pedales.

FOLIA inter digitos attrita odorem Afperulæ odoratæ
fpargunt, plerumque pubefcentia, fæpe leniter
tortuofa, membranâ ad bafin inftructâ, Vagina
ftriata, lævis.

SPICÆ. oblongo-ovatæ, laxæ.

CALYX: GLUMA bivalvis, Valvulis inæqualibus, inferiore dimido breviore, membranacea, acuta,
fuperiore acuminata, nervis tribus viridibus extantibus, *fig.* 3. 2.

COROLLA: GLUMA bivalvis, valvulæ fubæquales, membranaceæ, *pilofæ* ariftatæ, fufcæ; altera Arifta
quæ demum geniculata fit, prope bafin exfurgit,
altera prope apicem, *fig.* 4.

NECTARIUM: GLUMULÆ duæ, pellucidæ, nitidæ,
ovatæ, inæquales, *germen includentes,* fig. 5, 6.

STAMINA: FILAMENTA duo prælonga; ANTHERÆ
oblongæ, purpureæ, utrinque furcatæ, *fig.* 5.

PISTILLUM: GERMEN minimum oblongo-ovatum;
STYLI duo *filiformes* glumi longiores, verfus a-
picem plumulofæ, *fig.* 7.

SEMEN unicum, Nectario fufco, nitido, inclufum, *fig* .8.

ROOT perennial and fibrous.

STALKS at firft growing obliquely, finally becoming upright, ufually from 8 to 12 inches high.

LEAVES, if rubbed betwixt the fingers, fmelling like
Woodroff, generally pubefcent and often curled,
furnifhed with a membrane at bottom ; the
Sheath ftriated and fmooth.

SPIKES of an oblong oval fhape and fmooth.

CALYX : a GLUME of two Valves, the Valves unequal,
the lowermoft fhorter by one half, membranous
and acute; the uppermoft acuminated, with
three ftrong nerves or ribs, *fig.* 3. 2.

COROLLA : a GLUME of two Valves, the Valves nearly equal, membranous, hairy, of a brown colour,
and furnifhed with Ariftæ, one of the Ariftæ,
which finally becomes bent, fprings from the
bafe of the Valve, the other almoft at the top,
fig. 4.

NECTARIUM : two fmall, pellucid, fhining, oval, unequal Glumes or Valves *inclofing the Germen,*
fig. 5, 6.

STAMINA : two FILAMENTS very long: ANTHERÆ
long, purple, forked at each end, *fig.* 5.

PISTILLUM : GERMEN very fmall, of an oblong oval
fhape ; STYLES two, *flender,* longer than the
valves, and towards the top a little feathered,
fig. 7:

SEED fingle, inclofed within its brown, fhining Nectarium, *fig.* 8.

THE *Anthoxanthum* is diftinguifhed from the other Graffes by a very fingular circumftance, viz. that of having only
two *Stamina, fig.* 1. hence it is placed by LINNÆUS among the *Diandrous* plants, and feparated from all the other
Graffes ; this peculiarity, although it occafions a feparation which does violence as it were to Nature, yet it ferves in
a very ftriking manner to difcriminate this Genus from a numerous and difficult tribe of plants : exclufive of this fingularity, it differs alfo very effentially in the other parts of its fructification ; each of the Spiculæ contains in common
with many other graffes, only one flower, *fig.* 1 : one of the *Glume Calycinæ,* or valves of the Calyx, is fmall and
membranous, *fig.* 3 ; the other is large, and inclofes, or wraps up in it, as it were, the whole of the fructification,
fig. 3 ; thefe glumes, fo far as I have obferved, do not open and expand themfelves in the manner obfervable in the
Avena's, and many other graffes, were they feparate quite wide, and expofe their little feathery Styles ; but the Stamina
and Piftilla appear to pufh themfelves out, the glumes remaining clofed, *fig.* 1. The *Glumæ Corollaceæ* are very dif-
fimilar to thofe of moft other graffes, being remarkably hairy, and having each of them an Arifta, the longeft of which
fprings from near the bafe of the glume, is at firft ftraight, but as the feed becomes ripe, the top of it is generally bent
horizontally inward ; the other Arifta arifes from near the top of the oppofite Glume or Valve, *fig.* 4. The *Glumulæ Nectarii* or little Glumes of the Nectarium, differ no lefs in their ftructure, being compofed of two little oval
fhining Valves, one of which is fmaller than the other ; thefe clofely embrace the Germen, and cannot be feen but with
great difficulty, unlefs they are obferved juft at the time that the Antheræ are protruding from betwixt them, when
they are very diftinct, *fig.* 6 ; as foon as the Antheræ are excluded, they again clofe on the Germen, and continue to
form a coat to the feed which does not feparate. *fig.* 5, 8.

The Farmer, or thofe who have not been accuftomed to examine plants minutely, may readily diftinguifh this grafs
by its fmell ; if the leaves are rubbed betwixt the fingers, they impart a grateful odour like that of Woodruff,—hence
I have called it fweet-fcented.

Like the *Trifolium repens* or *Dutch Clover,* and many others of our moft ufeful plants, this Grafs grows on almoft
every kind of foil, from the pooreft and drieft, to the moft fertile and boggy ; it feems however in general to prefer a
foil that is moderately dry. It is fubject, like all other plants, to vary in its fize, according to the goodnefs of the
ground it grows in ; the leaves have a particular tendency to be curled if the foil be rich ; and when it grows in woods,
the fpikes are often much flenderer and loofer.

It has been called by fome Authors *Vernal* or *Spring Grafs,* from its coming into ear earlier than moft others ; towards the
middle of May it is in full bloom, and about the middle of June the feed is ripe—and may be eafily feparated on rub-
bing.

There is great reafon to believe, that this is one of our Graffes which might be cultivated with confiderable advantage : in the meadows about town it grows to a confiderable height, and forms a thick tuft of leaves at bottom ; but
the circumftance moft in its favour, is its early appearance in the Spring : this feems to point it out as a proper grafs
to fow with others in laying down meadow land, and probably the *Poa trivialis* or *common Meadow Grafs,* with the
Feftuca elatior or *Meadow Fefcue* joined to it, would form a mixture, the produce of which, would for this purpofe,
be fuperior to that of moft others.

Anthoxanthum odoratum

Aira aquatica.

AIRA *Linnæi Gen. Pl.* TRIANDRIA DIGYNIA.

Cal. 2 valvis, 2 florus. *Flofculi* abfque interjecto rudimento.

Raii Syn. Gen. 27. HERBÆ GRAMINIFOLIÆ FLORE IMPERFECTO CULMIFERÆ.

AIRA *aquatica* panicula patente, floribus muticis lævibus calyce longioribus, foliis planis. *Linnæi Syft.*

Vegetab. p. 96. *Fl. Suecic. No.* 68.

POA locuftis bifloris; glabris, florali gluma majori plicata, ferrata. *Haller hift. No.* 1471.

AIRA *aquatica Scopoli Fl. Carniol.* 94. *Hudfon Fl. Angl.* 29.

AIRA culmo inferiore repente, flofculis muticis calyce longioribus, altero pedunculato. *Roy. lugdb.* 60.

GRAMEN caninum fupinum paniculatum dulce. *Bauhin Pin.* 2.

GRAMEN miliaceum aquaticum. *Raii Syn.* 402. *Scheuz. agr.* 218.

GRAMEN miliaceum fluitans fuavis faporis. *Merret. Pin.*

RADIX perennis.
CULMUS bafi repit, furculofque emittit more Feftucæ fluitantis qui longe excurrunt et ad geniculos radiculas plures albas dimittunt; culmus demum erigitur, pedalis circiter, teres, erectus, fiftulofus, tener.

FOLIA latiufcula, tenera, lævia, carinata, vaginæ ftriatæ, ad bafin rubræ præcipue in furculis.

PANICULA erecta, diffufa, laxa, racemi plures ex uno puncto, fæpe flexuofi.

SPICULÆ plerumque bifloræ, flofculo uno feffili, altero pedunculato, purpurei, apicibus albidis, *fig.* 1.

CALYX: GLUMA bivalvis, valvulis inæqualibus, purpureis, lævibus, Corolla multo brevioribus, *fig.* 2.

COROLLA: GLUMA bivalvis, valvulis æqualibus, fubtruncatis, plicatis five angulatis, *fig.* 3.
STAMINA: FILAMENTA tria capillaria, longitudine Corollæ; ANTHERÆ flavæ, *fig.* 3.
PISTILLUM: GERMEN ovatum; STYLI duo plumofi, *fig.* 4.
NECTARIUM GLUMULÆ duæ minimæ ad bafin Germinis, *fig.* 5.
SEMEN ovatum, intra Glumas arcte claufum, *fig.* 7.

ROOT perennial.
STALK creeps at bottom, and fends out young fhoots like the Flote Fefcue grafs, which run out to a confiderable diftance, and fend down fmall white roots at the joints; it then becomes erect, grows to about a foot in height, is round, hollow, and tender.

LEAVES broadifh, tender, fmooth, carinated, the fheaths ftriated, red at bottom, particularly in the young fhoots.

PANICLE upright, fpreading, loofe; branches feveral, proceeding from one point, frequently crooked.

SPICULÆ generally contain two flowers, one of which is feffile, and the other ftands on a foot-ftalk, purple, the tips white, *fig.* 1.

CALYX: a GLUME of two valves the valves unequal, purple, fmooth, and much fhorter than the Corolla, *fig.* 2

COROLLA: a GLUME of two valves, the valves equal, as if cut off at top, folded or angular, *fig.* 3.
STAMINA: three capillary FILAMENTS the length of the Corolla; ANTHERÆ yellow, *fig.* 3.
PISTILLUM: GERMEN oval; STYLES two and feathery, *fig.* 4.
NECTARY two very minute GLUMES at the bottom of the Germen, *fig.* 5.
SEED oval, clofely contained within the Glumes, *fig.* 7.

The fame foil and fituation which produces the *Feftuca fluitans*, is productive alfo of this grafs; they both grow in gently flowing ftreams, or in wet boggy meadows; this circumftance may ferve among others to diftinguifh the *Aira aquatica* from fome of the *Poa's*, with which at firft fight the young botanift might eafily confound it: it has however befides this, many other characters which point it out more obvioufly. The bottom of the ftalk ufually creeps on the ground, and when it gets into the water, it runs out like the *Feftuca fluitans* to a confiderable diftance, throwing off roots and young fhoots as it paffes along, very much in the manner of that grafs: the ftalk grows about a foot or more in height, is hollow, and remarkably tender; the leaves are broader than any of the *Poa's*, except the *Poa aquatica*, which is in every refpect a much ftronger plant: but what more efpecially characterizes this grafs, is the purple or blueifh colour of the Panicles, which is difcernible even at a diftance; and the fweet tafte of the flowers if drawn through the mouth, whence this grafs has acquired the name of *dulce*. Its parts of fructification likewife above defcribed, diftinguifh it very ftrongly: when dried and placed between papers, the flowers and feeds are very apt to fall off.

It flowers in June and July, and may be found almoft every where in the fituations above-mentioned.

With refpect to its ufes in rural œconomy, it is in every refpect inferior to the *Flote fefcue grafs*, confequently not worth cultivating for the ufe of cattle.

In a country like ours, where cultivation has made a confiderable progrefs, the water plants are confined to a fmall fpace compared to what they occupied in a ftate of nature; the draining of bogs and lakes has rendered many large tracts in feveral parts of the kingdom, capable of producing corn and grafs adapted to the ufe of cattle, which were formerly inacceffible to man or beaft. We ought not however to look on this or any other plant as made in vain, becaufe we do not immediately fee the ufes they are applied to: feveral forts of water-fowl which abound in uninhabited countries, are expert gatherers of the feeds of the *aquatic graffes*; and no lefs than five different fpecies of *Mufci* or *Flies*, were produced from a few handfuls of the feeds of this grafs, which when I gathered it, were doubtlefs in their Pupa or Chryfalis ftate; How little do we know of natures productions!

Poa annua. Common dwarf Poa.

POA *Linnæi Gen. Plant.* TRIANDRIA DIGYNIA.

Raii Synop. Gen. 27. HERBÆ GRAMINIFOLIÆ FLORE IMPERFECTO CULMIFERÆ.

POA *annua*, panicula diffufa, angulis rectis, fpiculis obtufis, culmo obliquo compreffo. *Lin. Syft. Vegetab. p.* 97. *Spec. Plant. ed.* 3. *p.* 99. *Fl. Suecic. p.* 228.

POA culmo infracto, panicula triangulari, locuftis trifloris glabris. *Haller. hift. Vol.* 2. *p.* 223.

GRAMEN pratenfe paniculatum minus. *Baubin. Pin. p.* 2

GRAMEN pratenfe minimum album et rubrum. *Gerard. emac.* 5. *Parkinfon.* 1156.

GRAMEN pratenfa minus feu vulgatiffimum. *Raii Synop.* 408. *Hudfon. Fl. Angl. p.* 34. *Scopoli. Fl. Carniol.* 71. *Stillingfleet. tab.* 7

RADIX annua, fibrofiffima.

CULMI plures, cefpitofi, femiprocumbentes, in pratis vero inter alias plantas crefcentes, fuberecti, paululum infracti, femipedales.

VAGINÆ compreffæ, ancipites, læves.

FOLIA plurima, brevia, carinata, glabra, fæpe tranfverfim rugofa, margine minutiffime aculeata. *fig.* 8.

PANICULA triangularis, fubcompreffa, flores fubfecundi.

PEDUNCULI *univerfales* ad bafin paniculæ plerumque *bini*, altero breviore, in medio frequenter *terni*, apice vero *folitarii*; anguli nunc recti, nunc obliqui.

SPICULÆ ovato-acutæ, compreffæ, utrinque acutæ triflore, quadriflore. *fig.* 2.

CALYX: GLUMA bivalvis, valvulis concavis, inæqualibus. *fig.* 1.

COROLLA bivalvis, valvulis villofis, margine membranaceis, albidis, una majore, concava, obtufiufcula; altera minore, anguftiore. *fig.* 3.

STAMINA: FILAMENTA tria capillaria; ANTHERÆ flavefcentes, bifurcatæ. *fig.* 4.

PISTILLUM. GERMEN ovatum, STYLI duo ramofiffimi, pellucidi. *fig.* 5.

SEMEN ovatum, corolla adnafcente tectum, ad bafin villofulum. *fig.* 7.

ROOT annual and very fibrous.

STALKS numerous, forming a turf, femiprocumbent, but in meadows when growing among other plants, nearly upright, a little crooked, and about half a foot high.

SHEATHS flat, two edged, and fmooth.

LEAVES very numerous, fhort, keel-fhaped, fmooth, frequently wrinkled tranfverfely, the edge very finely ferrated. *fig.* 8.

PANICLE of a triangular fhape and flattifh, the flowers growing moftly to one fide.

PEDUNCLES: the *univerfal* peduncles generally proceed from the bottom of the panicle in *pairs*, one of which is fhorter than the other, from the middle often by *threes*, and at top *fingly*; forming angles fometimes ftraight, fometimes oblique.

SPICULÆ oval and pointed, flatifh and fharp on both fides, containing three and four flowers. *fig.* 2.

CALYX: a GLUME of two valves, the valves hollow and unequal. *fig.* 1.

COROLLA of two valves, the valves villous, membranous and whitifh at the edges, the one larger, hollow and bluntifh, the other fmaller and narrower. *fig.* 3.

STAMINA: the FILAMENTS very minute, the ANTHERÆ yellowifh and forked. *fig.* 4.

PISTILLUM: the GERMEN oval, two STYLES exceedingly ramified and pellucid. *fig.* 5.

SEED oval, covered by the Corolla which adheres to it, at bottom flightly villous. *fig.* 7.

THE laudable Society eftablifhed in London for the encouragement of Manufactures, Arts, and Commerce, fenfible of the improvements which might be made in Agriculture, from a more general introduction of the moft ufeful Englifh *Graffes*, have offered Premiums to fuch as fhall give the beft account of their cultivation, and the *Poa Annua* above defcribed, is one of thofe they have felected, from its appearing to them to be one of the moft ufeful.

Mr. Stillingfleet obferves that it makes the fineft turf, that he has feen in high Suffolk whole fields of it, without any mixture of other Graffes, and that as fome of the beft falt Butter we have in London comes from that County, he apprehends it to be the beft Grafs for the Dairy; from obferving likewife, that this Grafs flourifhed much more from being trodden on, he concludes that frequent rolling muft be very ferviceable to it.

There is no Grafs better entitled to Ray's epithet of *Vulgatiffimum* than this, as it occurs almoft every where, in Meadows, Gardens, at the fides of Paths, and on Walls: when it grows in a very dry fituation, it frequently doth not exceed three inches, but in rich meadows it often grows more than a foot in height. The panicle is frequently green, but in open fields it acquires a reddifh tinge; it flowers all the Summer long, and even in Winter if the weather be mild.

It appears to be the firft general covering which Nature has provided for a fruitful foil when it has been difturbed; for which reafon, in Walks, Pavements, or Pitching, it may be confidered as one of the moft troublefome of Weeds; the moft expeditious method of deftroying it, would probably be by pouring boiling water on it.

All the Authors that have defcribed this Grafs call it an annual, it differs however very confiderably from the other annual Graffes, they throw up their Spikes or Panicles, produce their flowers and feeds, and then die away; this on the contrary keeps continually throwing out new fhoots, and producing new flowers, and feeds, and if the ground be moift, a fingle plant will remain growing in this manner throughout the year, fo that we generally find on the fame plant, young fhoots and ripe feeds.

"*Hic ver affiduum atque alienis menfibus æftas.*"

Perhaps this is the only vegetable we have that in this Circumftance imitates the Tropical plants.

Although its feed may be gathered the whole fummer long, yet about the latter end of May, it will be found in the greateft plenty: Experience muft determine the beft method, in which this Grafs fhould be cultivated, whether by fowing its feed, or dividing and tranfplanting the Grafs itfelf; as this feed would with more difficulty be procured in large quantities than that of many others, and as a fingle tuft of this Grafs may be divided into a vaft number of young plants, probably tranfplanting it in wet weather would be the moft eligible mode of cultivation.

Thefe obfervations are fubmitted to the confideration of the Farmer and Gentlemen of landed property, who refide in the Country, and who have both leifure and opportunity to try experiments of this kind. Although the Authors province more particularly is to defcribe and figure thefe plants in fuch a manner as to make them as obvious as poffible, yet he would be exceedingly happy to communicate to the public, any improvements which may be made in this or any other branch of Agriculture, that he may be favoured with.

Festuca fluitans. Flote Fescue Grass.

FESTUCA *Linnæi Gen. Pl.* Triandria Digynia:

Raii Gen. 27. Herbæ Graminifoliæ flore imperfecto culmiferæ.

FESTUCA panicula ramofa erecta, fpiculis fubfeffilibus, teretibus muticis. *Linnæi Syft. Vegetab. p.* 107.

Fl. Suecic. p. 32.

POA locuftis teretibus multifloris, glumis floralibus exterioribus truncatis, interioribus bifidis. *Haller. hift. p.* 219. *n.* 1453. *v.* 2.

POA fluitans. *Scopoli Fl. Carniol. p.* 73.

GRAMEN aquaticum fluitans, multiplici fpica. *Bauhin Pin.* 2.

GRAMEN aquaticum cum longiffima panicula. *I. Baubin.* II. 490. *Raii Syn. p.* 412. Flote-Grafs.

GRAMEN fluviatile. *Gerard emac.* 14. *Parkinfon.* 1275. *Hudfon. Fl. Angl. p.* 38. *Oeder. Fl. Dan. t.* 237. *Schreber. Gram. tab.* 3. *Stillingfleet. mif. tab.* 10.

RADIX perennis, in limum profunde penetrans.

CULMUS pro ratione loci pedalis ad tripedalem, bafi repens furculofque promens, dein fuberectus, vaginis foliorum ad paniculam ufque amictus.

VAGINÆ foliorum compreffæ, fubancipites, ftriatæ.

FOLIA latiufcula, lævia; *furculorum* erecta, carinata, breviufcula, *caulina* longiora, planiufcula, flaccida, aquis tempore hyberno proftrata.

PANICULA longa, inclinata, nonnunquam fubfpicata fæpius vero ramofa, ramis nunc cauli adpreffis nunc diftantibus, ut pinxit Cl: Schreberus.

SPICULÆ tenues, teretes, unciales aut fefquicunciales 9 ad 12 floræ, rachi adpreffæ.

CALYX: Gluma bivalvis, valvulis inæqualibus, membranaceis. *fig.* 2.

COROLLA bivalvis, valvulæ longitudine æquales, calyce majores, *inferiore* majore, concava, lineata, nervis apice fæpe coloratis, apice membranacea, obtufiufcula, fæpius erofa; *fuperiori* lanceolata, compreffa, bicufpidata. *fig.* 3. 4.

STAMINA: Filamenta tria capillaria, Antheræ flavæ aut purpurafcentes, oblongæ, *fig.* 5.

PISTILLUM: Germen ovatum, Styli duo fubulati, reflexi, Stigmata ramofiffima. *fig.* 7. 6. 8.

NECTARIUM Glandula fquamiformis, cordata, horizontalis, ad bafin germinis. *fig.* 9.

SEMEN oblongum, nitidum olivaceum, bicorniculatum, nudum. *fig.* 10. 11.

FIG 12 Spicula morbo *Ergot* affecta.

ROOT perennial, ftriking deep into the mud.

STALK according to its place of growth from one to three feet in length, creeping at bottom and fending forth young fhoots, afterwards nearly upright; covered with the fheaths of the leaves as far as the panicle.

SHEATHS of the leaves, flattened, two edged, and ftriated.

LEAVES rather broad and fmooth, thofe of the young fhoots upright, keel-fhaped, and fhortifh; thofe of the ftalk longer, flattifh, weak, and hanging down, in the winter feafon lying flat on the water.

PANICLE long, generally inclined or bending down a little, fometimes forming a kind of fpike, but moft commonly branched; the branches fometimes preffed to the ftalk, fometimes diverging from it in the manner reprefented by Schreber.

SPICULÆ flender, round, an inch or an inch and a half long, producing from 9 to 12 flowers, preffed to the Stalk.

CALYX: a Glume of two valves, which are unequal and membranous. *fig.*2.

COROLLA of two valves, which are of an equal length and bigger than the Calyx, the *lower valve* largeft, concave and nervous, the nerves towards the top frequently coloured, at top membranous, rather blunt with uneven points, the upper valve more pointed, flat and bifid. *fig.* 3. 4.

STAMINA: three Filaments very flender, Antheræ oblong and yellow or purplifh. *fig.* 5.

PISTILLUM: Germen oval, Styles two, tapering and bending back, Stigmata very much branched. *fig.* 7. 6. 8.

NECTARY a fmall heart-fhaped fquamiform gland, placed horizontally at the bottom of the Germen. *fig.* 9.

SEED oblong, fhining, of an olive colour, with two little horns, and naked. *fig.* 10. 11.

FIG 12 a Spicula affected with the difeafe called *Ergot*.

IN fpeaking of the *Bromus mollis*, we had occafion to remark the very great variety of appearance to which the Graffes were fubject from foil and fituation, and this obfervation is equally applicable to the *Feftuca fluitans*.

This Grafs appears to thrive beft in ftill waters, or gently running ftreams, where its numerous fibres penetrate eafily into the mud; in fuch fituations it becomes very luxuriant, the leaves are large, tender and fweet, and the Panicle becomes very much branched; but in Meadows where it is deprived of its natural quantity of water, it becomes in every refpect lefs, and the Panicle is frequently changed to a fimple fpike: when it has nearly done flowering, the branches of the Panicle generally project from the main ftalk fo as to form an acute angle. In every fituation whether the Panicle be large, or fmall, the Spiculæ are always preffed clofe to the ftalk or branches of the Panicle, and this circumftance joined to the length, and roundnefs of the Spiculæ, fufficiently characterize this fpecies; if it fhould not however, its parts of fructification afford at once a moft pleafing and fatisfactory diftinction, *vid. fig.* 6. 9. 10.

We

Festuca fluitans

We have often had the singular pleasure of observing this Grass soon after being gathered, expand its Glumes and expose its delicate yellow Stamina, and still more delicate Pistilla, and in this expanded state each Spicula puts on a very different face, and seems to invite the Student to its investigation, and would he wish to become acquainted with the structure of this useful tribe of plants, he cannot select one more proper for his purpose, as it may be found in almost every watery ditch, flowering from the beginning to the end of Summer, and has all the parts of fructification which are peculiar to the Grasses, large enough to be distinctly distensed even by the naked Eye, and so exposed as to be visible without the trouble of dissection.

Modern Botanists seem much divided whether they should consider this as a *Poa* or *Festuca*, as it does not appear to us that we should in the least advance our favourite Science by altering its generic name, we have continued that of LIN-NÆUS, although we are by no means satisfied with his generic characters of the Grasses in general, and are persuaded that future observations and a more accurate attention to the minute parts of their fructification, will place those Genera in a much clearer point of view than has yet been done by any author.

Professor OEDER in his FLORA DANICA, and the celebrated SCHRÆBER in his AGROSTOGRAPHIA, have both given a figure of this grass. As we have not seen it growing either in Denmark or Germany we cannot say that their figures do not express its particular mode of growth in those countries, but they do not convey to us its habit or manner of growing here; in both their figures the Panicle is represented quite upright, whereas with us it is always more or less inclined; this however is a matter of no great moment, a deviation from nature in the representation of the minute parts of the fructification is a matter of much greater consequence, and we are sorry to find that Mr. SCHRÆBER whose knowledge and accuracy can seldom be called in question, has not been sufficiently attentive to all the parts which characterize this species. He has represented the Styles as branched or feathered quite down to the German, whereas they are evidently *naked at bottom* and *much branched at top only*; the singular Squamula or Scale at the base of the German he has properly noticed, but the *two little Horns* at the top of the seed, which are the remains of the Styles, and which in a peculiar manner distinguish this important seed, he does not remark. In the *Flora Danica* the Styles are likewise feathered down to the German and the Squamula at the base of the German wholly omitted.

This Grass is found to be of considerable importance in the œconomy of Nature.

The *Phalæna Festucæ* or Gold Spot Moth, to which LINNÆUS with great propriety adds the epithet of *pulcherrima*, (*vid. Fauna Suecica. p.* 311. *Albin. pl.* 84 *lit.* E. F. G. H.) is said by him to feed on this particular Species; with us however it is always found on a different grass, viz. the *Poa aquatica* or *large water Poa*; its history, with the particular manner of finding it we shall give under that grass.

From the observations of late writers, it appears that several sorts of Cattle are remarkably fond of this grass, particularly Kine and Hogs, and that in the spring time they are frequently enticed into bogs by endeavouring to get at its sweet young shoots, which appear earlier than those of most other Grasses.

" Professor KALM in a journey through part of *Sweden*, observed the Swine to go a great way into the water after
" this grass, the leaves of which, they eat with great eagerness; on this he was tempted to try if they would eat the
" same grass dried; he accordingly had small bundles of it gathered, dried, and cast before them; the consequence
" was they ate it seemingly with as much appetite as horses do hay, hence he concludes that by cultivating this grass,
" *wet and swampy places* might be rendered useful, and a great deal of corn, &c. saved".

He who introduced the method of feeding hogs in summer time on Clover, deserved very well of his country; and if the hay of this grass would keep them in heart during the winter, it might prove a very valuable discovery.

Mr. Kent in his *hints to Gentlemen of landed property*, lately published, considers this as a most valuable grass, and assures us (*p.* 34) it is to be improved above all others, and at a less expence, merely by flooding; (*p.* 54,) he informs us that flooding destroys *all weeds*, and enriches the land to a very high degree; (*p.* 56,) he says as rolling and pressure *bring* the *annual meadow-grass*, so flooding immediately begets the *flote fescue*. These assertions of Mr. Kent bespeak neither the Philosopher nor the accurately practical Farmer, they contain an exaggerated account of improving pasture land by a particular process, but show a great want of that minute attention which so important a subject required.

From a long residence in Hampshire, we well know that the meadows in that county are considerably improved by flooding them, that is stopping the water when there happens to be an unusual quantity from violent or long continued rains, and by means of trenches or gripes, conveying the surplus water so as to overflow them entirely if possible; but we deny, that by this process *all weeds* are destroyed, the use of *manure* superseded, or that *flote fescue grass* is immediately *begotten*. Although it is a constant practice with the farmers to flood their meadows in the winter, it is no less a constant practice with such as wish to have good crops of grass to manure them with dung or ashes. Flooding can no otherwise destroy weeds than by altering the soil in which they grow, and if it destroys one set of weeds, it must certainly favour the growth of another: if those plants which thrive best in a dry situation are destroyed by the alteration which now takes place in the soil, those which are fond of a moist situation will proportionably flourish. If the *flote fescue grass* was immediately produced by flooding, we should find all those meadows which have undergone this operation to contain nothing but this kind of grass, whereas the richest and best meadows in Hampshire contain scarce a single blade of it: the fact is, this grass will not flourish in meadow land, unless you convert it into a kind of bog or swamp, and I believe few landed Gentlemen will think this an improvement, or thank Mr. Kent for giving them such a hint.

" Mr. Stillingfleet informs us that Mr. Deane a very sensible Farmer at Ruscomb, in Berkshire, assured him, that
" a field always lying under water of about four acres, that was occupied by his father when he was a boy, was covered
" with a kind of grass that maintained five farm-horses in good heart from April to the end of harvest without giving
" them any other food, and that it yielded more than they could eat. He at my desire brought me some of the
" grass, which proved to be the *flote fescue* with a mixture of *marsh bent*; whether this last contributes much towards
" furnishing so good pasture for horses I cannot say, they both throw out roots at the joints of the stalks and therefore
" likely to grow to a great length. In the index of dubious plants at the end of *Ray's Synopsis*, there is mention
" made of a grass under the name of *Gramen caninum supinum longissimum* growing not far from *Salisbury* twenty-four feet
" long; this must by its length be a grass with a creeping stalk; and that there is a grass in Wiltshire, growing in
" watery meadows, so valuable that an acre of it lets from ten to twelve pounds, I have been informed by several
" persons. These circumstances incline me to think it must be the *flote fescue*; but whatsoever grass it be it certain-
" ly must deserve to be enquired after".

It may not be improper to add, that the account of the extraordinary long grass above mentioned, was taken by RAY from the *Phytographia Britannica*, which mentions the particular spot where it grew, viz. at Mr. Tucker's, at Maddington, nine miles from Salisbury; it is also remarked that *they fat Hogs with it*.

As it is now above a century since this enquiry was first made, is it not surprizing that no succeeding Botanic Writer should have acquired satisfactory information concerning it? I am promised specimens of the roots and seeds.

Upon

Upon the whole, from the obfervations which we ourfelves have made on this Grafs and from what is to be collected from Authors, it appears that if it be cultivated to any advantage it muſt be in ſuch meadows as are *naturally very wet and never drained*.

The quickeſt and perhaps the beſt method of propagating it would be by tranſplanting the roots at a proper ſeaſon, and if the ſoil prove ſuitable, from the quickneſs of its growth, and its creeping Stalk, it would ſoon exclude moſt other plants, and produce a plentiful crop.

In foreign countries the ſeed of this Grafs ſeems to be an object of more importance than the grafs itſelf, the following is the ſubſtance of what Mr. Schreber has ſaid concerning it, (vid. *Beſchreibung der Gräſer p. 40.*)
" The ſeed has a ſweet and pleaſant taſte particularly before it comes to its full growth, whence the plant has
" acquired the name of *Manna Grafs*. Ducks and other water-fowl feed on it with much eagerneſs (Linnæus has
" remarked that the Water-fowl are very well acquainted with the method of collecting theſe ſeeds) it has been
" obſerved likewiſe that Fiſh are fond of it, and that Trout in particular thrive in thoſe rivers where this grafs grows
" in plenty and ſheds its ſeeds; but it is not only for Birds and Fiſh but alſo for Man a palatable and nutritious
" food, and has for many years paſt been known at Gentlemens tables under the name of *Manna-Grout*.

" The Manna Grafs is of two kinds the one *Panicum ſanguinale* or *Cocks-foot Panic-Grafs* the other *Feſtuca fluitans*
" which we have now deſcribed; the former is cultivated in ſeveral parts of Germany, and its ſeed ſomewhat reſem-
" bles that of Millet, the latter is collected in great abundance from the plant as it grows wild in *Poland*, *Lithuania*,
" the *new Marche* and about *Franckfort* and other places in *Sileſia* as alſo in *Denmark* and *Sweden* and hence exported
" to all parts.

" The common method they make uſe of to gather and prepare this ſeed in *Poland*, *Pruffia*, and the *Marche* is
" as follows. At ſun riſe the ſeed is gathered or beat from the dewy grafs into a horſe-hair ſieve, and when a
" tolerable quantity is collected, it is ſpread on a ſheet and dried fourteen days in the ſun; it is then thrown into
" a kind of wooden trough or mortar, ſtraw or reeds laid between it, and beat gently with a wooden Peſtle ſo as to
" take off the chaff and then winnowed. After this it is again put into the mortar, in rows, with dried Marygold-
" flowers, Apple, and Hazel leaves, and pounded until the Hufk is entirely ſeparated and the ſeed appears bright,
" it is then winnowed again, and when it is by this laſt proceſs made perfectly clean it is fit for uſe. The Mary-
" golds are added with a view to give the ſeeds a finer colour. The moſt proper time for collecting them is in July.
" A Buſhel of the ſeed and chaff, yields about two quarts of clean ſeed.

" When boiled with milk or wine they form an extremely palatable food, and are moſt commonly made uſe of
" whole in the manner of *Sago* to which they are in general preferred.

In the month of *October* laſt, I diſcovered in a watery ditch, which runs through a meadow not far from Kent-
Street Road an uncommon appearance in ſome of the ſeeds of this grafs, and on a farther examination, I found
whole Panicles the ſeeds of which were affected in a ſimilar manner, inſtead of being of their natural ſize, and colour,
they were enlarged to a very great degree, affumed externally a blackiſh colour, and were more or leſs incurvated.
Struck with the novelty as well as oddity of the appearance I conjectured at firſt that it was a diſeaſe occaſioned by
ſome Infect, I examined it more attentively, but could not find the leaſt cauſe to ſuppoſe that an Infect had been
concerned in it. The ſurface of ſome of theſe ſeeds was ſoft, and chopped, they were light as to weight, internally of a whitiſh colour, inſiped in their taſte but not diſagreeable. Having a little before this been favoured with
a ſight of ſome *horned Rie* it now occurred to me that this was the ſame diſeaſe which had been ſaid to affect the
Rie only, and farther enquiry confirmed my conjecture.

As this ſingular diſeaſe of the Rie has firſt been noticed by the French, and as ſome very uncommon circum-
ſtances have attended it, it cannot fail of proving acceptable to our readers to lay before them the ſubſtance of
what they have ſaid concerning it. In the *Hiſtoire de L'Academie royale des Sciences* there is an account given of a
particular ſpecies of Gangrene or Mortification which attacked many perſons in ſome particular provinces of France.
" It began generally at the toes and ſometimes ſpread as high as the thigh. Out of fifty people there was but
" one that was attacked with this diſeaſe in the hands and what was equally remarkable there were no females
" affected with it except ſome little Girls,

" It appears that this ſingular malady attacked only the lower ſort of people, and that too in years of ſcarcity,
" that it proceeded from bad nouriſhment, and principally from eating bread made of a certain black and diſeaſed
" corn called Ergot, from the grains affuming ſomewhat of the form of a Cocks Spur. *vid. fig.* 12.

" The manner in which this ſingular monſtroſity of the Corn is produced is thus related by Monſieur Fagon.

" There are certain miſts which prove injurious to the corn, and from which the greateſt part of the Ears of the
" Rie defend themſelves by their beards, In thoſe however which this hurtful humidity can ſtrike and penetrate,
" it rots the ſkin which covers the grain, blackens it, and alters the ſubſtance of the grain itſelf, the juices which
" form the ſeed being no longer kept within their ordinary bounds by the ſkin, are carried hither in two great an
" abundance and amaſſing themſelves irregularly form this monſtrous appearance.

" He obſerves that it is *only in Rie* that the *Ergot* is to be found, that the poor people do not ſeparate this grain
" from that which is good, that it was only in ſuch particular ſeaſons as favoured the growth of the *Ergot* that
" this diſeaſe was prevalent, that the country people after eating bread made of this bad corn perceived themſelves
" as if drunk, and after this the mortification generally took place, that in ſome provinces were there was but little
" of this Ergot this ſpecies of diſeaſe was not known.

" From the obſervations made by the Farmers of that country it appears that this bad ſpecies of grain is pro-
" duced in the greateſt abundance in ſuch land as is wet and cold, and particularly in rainy ſeaſons. The Poultry
" refuſed it when given them, neverthelefs if by accident they had eaten it they did not appear to be hurt by it.
" When ſown (as might be expected) it did not vegetate. "

A kind of mortification very ſimilar to the above deſcribed was obſerved in this Kingdom ſome years ago; it
affected the ſame kind of people and on enquiry it was found that they had fared very hard, and that the bread
which they had eaten was made of the *tailings* or *ſcreenings* of Corn, but it was not aſcertained whether it contained
any of the Ergot or not.

From the inſipid taſte of this corn, as well as from its not proving fatal to Poultry, It ſeems exceedingly probable
that it is not in itſelf noxious, any otherwiſe than as it affords no nouriſhment; and that thoſe people who have
eaten of this corn, have in fact been abridged of a proportionate quantity of food, hence from an impoveriſhed
ſtate of the fluids and a weak action of the veſſels this ſpecies of Mortification might eaſily be induced.

BROMUS MOLLIS. SOFT BROME GRASS.

BROMUS   *Linnæi Gen. Pl.* TRIANDRIA DIGYNIA.

        Raii Syn. Gen. 27. HERBÆ GRAMINIFOLIÆ FLORE IMPERFECTO CULMIFERÆ.

BROMUS   *mollis* panicula erectiufcula, fpiculis ovatis pubefcentibus, ariftis rectis, foliis molliffime villofis

        Linnæi Syft. Vegetab. p. 102. *Sp. Pl. p.* 112.

BROMUS   hirfutus, locuftis feptifloris, ovato conicis. *Haller hift. p.* 1504.

BROMUS   Polymorphus. *Scopoli Fl. Carniol. p.* 80.

FESTUCA   avenacea hirfuta, pauiculis minus fparfis. *Raii Synop. p.* 413   *Hudfon Fl. Angl. p.* 39. *n.* 1.

        Secalinus. *Schreber. Gram. pl.* 6. *fig.* 1.

RADIX biennis *	ROOT biennial *
CULMUS pedalis ad tripedalem, erectus; GENICULI tumidi, cylindracei.	STALK from one to three feet high, upright; the JOINTS fwelled and cylindrical.
FOLIA cum VAGINIS pilis mollibus veftita.	LEAVES together with their SHEATHS covered with foft hairs.
PANICULA erectiufcula, nunc coarctata nunc diffufa.	PANICLE nearly upright, fometimes clofe, fometimes fpreading.
SPICULÆ *ovato-acutæ*, turgidæ, fubcompreffæ, plerumque villofæ, octofloræ, circa oras glumarum albidæ. *fig.* 1.	SPICULÆ. *oval and pointed*, turgid, flattifh, generally villous, containing eight flowers, whitifh about the edges of the Glumes. *fig.* 1.
CALYX: GLUMA bivalvis, valvulis inæqualibus, muticis. *fig.* 2.	CALYX: a GLUME of two valves, the valves unequal without any beard, or arifta, *fig.* 2.
COROLLA: GLUMA bivalvis, valvulâ exteriore lata, concava, ftriata, ariftata, *fig.* 4. interiore planiufcula, *ciliata*, lanceolata. *fig.* 3. ARISTA valvulis paulo longior, fubrecta, *fig.* 4.	COROLLA: a GLUME of two valves, the outermoft valve broad, hollow, ftriated, and bearded, *fig.* 4; the innermoft flattifh, *ciliated or hairy at the edges* and pointed, *fig.* 3; the ARISTA a little longer than the valves and nearly ftraight, *fig.* 4.
NECTARIUM: Glumula bipartita, ad bafin petali interioris, *fig.* 5, parum auct:	NECTARIUM: a fmall kind of Glume deeply divided, placed at the bafe of the inner petal, *fig.* 5. a little magnified.
STAMINA: FILAMENTA tria capillaria, ANTHERÆ primum flavæ, oblongæ, dein fufcæ et bifurcatæ. *fig.* 7. 6. auct:	STAMINA: three FILAMENTS very fmall, ANTHERÆ firft yellow and oblong, laftly brown and forked at each end, *fig.* 7. 6. magnified.
PISTILLUM: GERMEN ovatum, apice fubemarginatum, *fig.* 8. STYLI duo, ad bafin ufque plumofi, *ex uno latere germinis enati. fig.* 9.	PISTILLUM: GERMEN oval, with a flight depreffion at top, *fig.* 8. two STYLES feathery quite down to the bottom, *proceeding from one fide of the Germen, fig.* 9.
SEMEN oblongum, concavum, calyci adnatum *fig.* 10. denudatum *fig.* 11.	SEED oblong, concave, adhering to the Calyx *fig.* 10. the Calyx taken off, *fig.* 11.

OUR Farmers in general are not very warm in their recommendations of this Grafs, neverthelefs it abounds in moft of our beft meadows; it fprings up early, and ripens its feed generally about the time of Hay-making. The feed is large, and each panicle contains nearly as much as that of a common Oat, indeed it feems to have more pretenfions to the the name of Corn than of Grafs.

Although Cattle may not be fo fond of the leaves, and panicle of this Grafs while green as of fome others, yet may it not (when cut down as it ufually is when the feed is nearly ripe) contribute to render the hay more nutritive? and hence may it not be a proper Grafs to fow with others.? It feems at leaft to deferve the attention of the Farmer.

There is perhaps no clafs of plants more affected by difference of foil and fituation than the Graffes, hence the fame plant has often been divided into feveral fpecies; and to fuch Varieties is the prefent Plant incident, as to occafion SCOPOLI to give it the name of *Polymorphus*.

When it grows on a Wall, or dry Bank, the Spiculæ are generally more upright, and clofer together; when the foil is rich and moift, the Spiculæ fpread out, and the whole plant becomes much larger; iu Meadows the Spiculæ frequently lofe their villous appearance and become perfectly fmooth. To determine this fpecies then with more certainty, recourfe muft be had to the parts of fructification.

* According to Linnæus

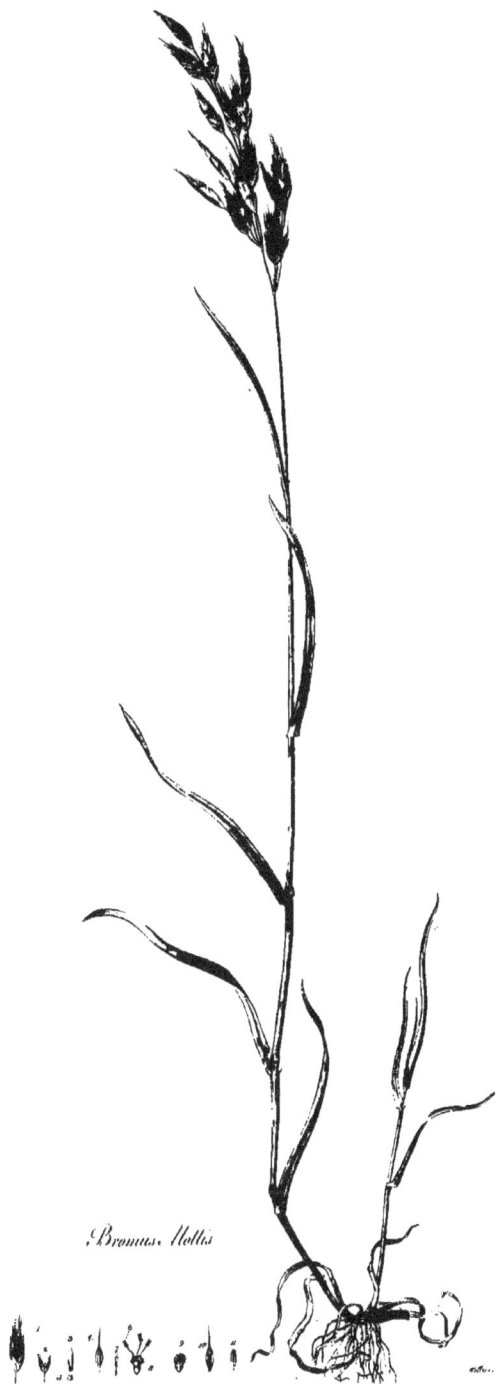

Bromus Mollis

Bromus sterilis

Bromus Sterilis. Barren Brome Grass.

BROMUS *Linnæi. Gen. Pl.* Triandria Digynia.

Raii Gen. 27. Herbæ graminifoliæ, flore imperfecto culmiferæ.

BROMUS *sterilis*, panicula patula, spiculis oblongis distichis, glumis subulato-aristatis. *Lin. Syst. Vegetab. p.* 103.

BROMUS panicula nutante; locustis septifloris; glumis argute lanceolatis, lineatis, subhirsutis. *Haller. hist. n.* 1505.

FESTUCA avenacea sterilis elatior. *Bauhin. pin.* 9. 10.

BROMOS herba, five avena sterilis. *Parkinson,* 1147. Bromos sterilis. *Gerard. emac. Raii Synop. p.* 412. Great wild Oat-Grass or Drank. *Hudson. Fl. Angl. p.* 40. *Scopoli. Fl. Carniol. p.* 78.

RADIX fibrosa.

CULMI pedales ad bipedales, suberecti, teretes, læves, ad basin infracti; Geniculi tumidi.

FOLIA longa, plana, unâ cum vaginis mollissime villosa.

PANICULA magna, nutans; Pedunculi plerumque simplices, *ad basin tumidi.*

SPICULÆ biunciales, subcompressæ, apice divergentes, *fig.* 1.

CALYX: Gluma bivalvis, Valvulis inæqualibus lineari-lanceolatis, *fig.* 2.

COROLLA: bivalvis, Valvulis inæqualibus, exteriore longiore, concava, striata, apice membranacea, bifida, Arista recta Corollâ duplo longiore terminata, *fig.* 3. Valvulâ interiore planiuscula, ciliata, *fig.* 4.

NECTARIUM: Glumulæ duæ acuminatæ, ad basin biglandulosæ, *fig.* 5.

STAMINA: Filamenta tria, capillaria, Antheræ flavæ, *fig.* 5.

PISTILLUM: Germen oblongum, apice truncatum sive emarginatum, pars inferior ex quâ styli prodeunt et quod verum Germen esse videtur, nitida, *fig.* 7. pars superior albida, villosa, *fig.* 8. Styli duo plumosi, patuli, *fig.* 9.

SEMEN ex purpureo-fuscum, oblongum, aristatum, calyce tectum, *fig.* 10, denudatum, *fig.* 11.

ROOT fibrous.

STALKS from one to two feet high, nearly upright, round and smooth, at bottom crooked or elbowed; the Joints swelled.

LEAVES long and flat, covered, together with their sheaths with soft short hairs.

PANICLE large, and drooping, the Peduncles generally simple, and *swelled at their base.*

SPICULÆ about two inches long, flattish and diverging toward the extremity, *fig.* 1.

CALYX: a Glume of two Valves, the valves inequal, long and narrow, *fig.* 2.

COROLLA: composed of two Valves, which are inequal, the exterior Valve longest, concave, striated, at top membranous and bifid, terminated by a straight Arista twice the length of the Corolla, *fig.* 3. the interior Valve nearly flat, and ciliated, *fig.* 4.

NECTARY: two small long-pointed glumes with a small gland at the base of each, *fig.* 6.

STAMINA: three small Filaments: the Antheræ yellow, *fig.* 5.

PISTILLUM: the Germen oblong, at top flat or slightly emarginate, the bottom part from whence the Styles proceed, and which seems to be the true Germen, is smooth and shining, *fig.* 7. the upper part white and villous, *fig.* 8. two Styles, feathery and spreading, *fig.* 9.

SEED of a purplish brown colour, oblong, bearded, enclosed within the Calyx, *fig.* 10. the Calyx stripped off, *fig.* 11.

Much praise is due to the late ingenious Mr. Stillingfleet for his attempts to introduce, more generally among Farmers, a knowledge of the most useful English Grasses: his observations on this subject are so exceedingly pertinent that the insertion of them cannot fail to prove highly acceptable to such as have the promotion of Agriculture at heart.

"It is wonderfull to see how long mankind has neglected to make a proper advantage of plants of such importance, " and which in almost every country are the chief food of cattle. The farmer for want of distinguishing, and selecting " grasses for seed, fills his pasture either with weeds, or bad, or improper grasses; when by making a right choice, " after some trials he might be sure of the best grass, and in the greatest abundance that his land admits of. At present " if a farmer wants to lay down his land to grass, what does he do? he either takes his seeds indiscriminately from " his own foul hay-rick, or sends to his next neighbour for a supply. By this means, besides a certain mixture of all " sorts of rubbish, which must necessarily happen; if he chances to have a large proportion of good seeds, it is not " unlikely, but that what he intends for dry land may come from moist, where it grows naturally, and the contrary. " This is such a slovenly method of proceeding, as one would think could not possibly prevail universally; yet this " is the case as to all grasses except the darnel grass, and what is known in some few countries by the name of the " Suffolk grass; and this latter instance is owing, I believe, more to the soil than any care of the husbandman. " Now would the farmer be at the pains of separating once in his life half a pint, or a pint of the different kinds of " grass seeds, and take care to sow them separately; in a very little time he would have wherewithal to stock his " farm properly, according to the nature of each soil, and might at the same time spread these seeds separately over " the nation by supplying the seed-shops. The number of grasses fit for the farmer is, I believe small; perhaps half " a dozen, or half a score are all he need to cultivate; and how small the trouble would be of such a task, and how " great the benefit, must be obvious to every one at first sight. Would not any one be looked on as wild who should " sow *wheat, barley, oats, rye, pease, beans, vetches, buck-wheat, turnips* and weeds of all sorts together? yet how is it " much less absurd to do what is equivalent in relation to grasses? does it not import the farmer to have good hay " and grass in plenty? and will cattle thrive equally on all sorts of food? we know the contrary. Horses will " scarcely eat hay, that will do well enough for oxen and cows. Sheep are particularly fond of one sort of grass, " and fatten upon it faster, than on any other in Sweden, if we may give credit to Linnæus. And may they not do " the same in England? How shall we know till we have tried? Nor can we say that what is valuable in Sweden " may be inferior to many other grasses in England; since it appears by the Flora Suecica that they have all the " good ones that we have. But however this may be I should rather chuse to make experiments, than conjectures."

The present Grass is not one of those which are worth the Farmer's cultivation, but so much the reverse, that most Authors have called it *sterilis*, not because it is really barren but from its inutility with respect to Cattle.

It grows exceeding common under hedges and flowers in May and June.

In order to have a clear idea of the structure of the parts of fructification in the Grasses, they should be examined just at the time, or rather before the *Antheræ* have discharged their *Pollen*, a small space of time makes a considerable alteration in their appearance.

In this species of *Bromus* as well as in the *Bromus mollis* the Styles proceed from the *middle* of the Germen and not from the *top*, this is a peculiarity which seems to have escaped the notice of *Schreber* who has written professedly on the Grasses, and examined them with more accuracy than any preceding Writer. In his figures the *Styles* proceed always from the Apex of the *Germen.*

Dipsacus pilosus.

Dipsacus pilosus. Small wild Teasel or Shepherd's Rod.

DIPSACUS *Linnæi Gen. Pl.* Tetrandria Monogynia.

Calyx communis polyphyllus ; proprius fuperus. *Receptaculum* paleaceum.

Raii Syn. p. 191. Herbæ Corymbiferis affines.

DIPSACUS *pilofus* foliis petiolatis appendiculatis. *Linn. Syft. Vegetab. p.* 120. *Spec. Plant.* 141.

DIPSACUS foliis biauribus, capitulis hemifphæricis. *Haller. hift. helv.* No. 199.

DIPSACUS fylveftris capitulo minore vel virga paftoris minor. *Bauhin Pin. p.* 385.

DIPSACUS minor feu Virga paftoris. *Ger. emac.* 1168.

VIRGA PASTORIS. *Parkinfon* 984. *Raii Synop. p.* 192. *Hudfon. Fl. Angl. p.* 49.

RADIX biennis.

ROOT biennial.

CAULIS orgyalis, erectus, ramofiffimus, pene teres, aculeatus, fulcatus.

STALK about fix feet high, upright, very much branched, nearly round, prickly and grooved.

RAMI oppofiti, patentes, cauli fimiles.

BRANCHES oppofite, fpreading, like the ftalk.

FOLIA ad *bafin* Caulis, connata, ovato-lanceolata, ferrata, nervo medio fubtus aculeato, *indivifa*, *fuprema* appendiculata ; ramorum ; *ima* appendiculata, ferrata, *fuprema* margine integerrima, lanceolata.

LEAVES *at the bottom of the* Stalk connate, ovato-lanceolate, ferrated, the midrib prickly underneath, *undivided*, thofe at the *top* dividing at the bafe into two fmaller leaves ; the leaves on *the branches* at *bottom* fimilar to thofe laft defcribed, at *top* lanceolate, with the edges entire.

PEDUNCULI erecti, longi, ex dichotomiâ caulis, fulcati, aculeati, apice fpinofiffimi, uniflori.

FOOT-STALKS of the flowers, upright, long, proceeding from the middle where the ftalks feparate, grooved, prickly, at top very full of flender fpines, fupporting one flower.

FLORES albidi, in capitulum hemifphæricum collecti, dum florent nutantes, poftea capitula eriguntur.

FLOWERS whitifh, collected together in a fmall hemifpherical head, which, while the plant is in flower, droops, and afterwards becomes upright.

CALYX: Perianthium *commune* multiflorum, hexaphyllum, foliolis longitudine florum, patentibus, lanceolatis, mucronatis, *fig.* 1 : Perianthium *proprium* parvum, fuperum, concavum, ciliatum, *fig.* 5. lente auctum.

CALYX : the *common* Perianthium fupporting many flowers, compofed of fix leaves, the length of the flowers, fpreading, lanceolate and pointed, *fig.* 1. The Perianthium of *each flofcule* fmall, placed above the Germen, hollow, and ciliated, *fig.* 5, magnified.

COROLLA *propria* monopetala, tubulofa, limbo qudrifido, laciniâ inferiore longiore, *fig.* 3.

COROLLA : *each flofcule* monopetalous, tubular, the limb quadrifid, the lowermoft fegment longeft, *fig.* 3.

STAMINA : Filamenta quatuor Corollâ longiora ; Antheræ purpureæ, *fig.* 3.

STAMINA : four Filaments, longer than the Corolla ; Antheræ purple, *fig.* 3.

PISTILLUM: Germen inferum, tetragonum ; Stylus filiformis, longitudine Corollæ ; Stigma fimplex, *fig.* 6.

PISTILLUM : Germen placed below the Calyx, quadrangular ; the Style filiform, the length of Corolla ; the Stigma fimple, *fig.* 6.

PERICARPIUM nullum.

SEED-VESSEL wanting.

SEMINA fufca, fubtetragona. *fig.* 4.

SEEDS brown, nearly quadrangular.

RECEPTACULUM commune hemifphæricum, paleaceum, pars inferior palearum concava, alba, carinata, fuperior lanceolata, acuminata, fpinulis obfita. *fig.* 2.

RECEPTACLE common to all the flowers paleaceous : the lower part of the paleæ hollow, white, and angular behind ; the upper part lanceolate, tapering to a point, and befet with little fpines or hairs, *fig.* 2.

THIS fpecies of *Teafel* may be confidered as one of our Plantæ rariores; hitherto I have found it only in one place near town, viz. on the right hand fide of the Turnpike-road leading from *Deptford* to *Lewifham*, not far from the latter : as it grows to a confiderable height, it is confpicuous at a diftance : the flowers appear in July, and the feed is ripe in September : it continues to blow for a confiderable time, and did not the plant take up fo much room, there is beauty enough in its flowers to recommend it for the Garden. Moths feem very fond of its bloffoms, being found on them in great numbers after fun-fet.

HOTTONIA PALUSTRIS. WATER HOTTONIA, OR WATER VIOLET.

HOTTONIA *Lin. Gen. Pl.* PENTANDRIA MONOGYNIA.

Raii Syn. Gen. 18. HERBÆ FRUCTU SICCO SINGULARI, FLORE MONOPETALO.

HOTTONIA *palustris*, pedunculis verticillato-multifloris. *Lin. Syst. Vegetab.* 164.

HOTTONIA florum verticillis spicatis. *Haller. hist. n.* 632.

MILLEFOLIUM aquaticum seu Viola aquatica, caule nudo. *Bauhin. pin.* 141. *Parkinson,* 1256.

VIOLA palustris. *Gerard. emac.* 826. *Raii Syn. p.* 285. *Hudson. Fl. Angl. p.* 72. *Scopoli Fl. Carniol. n.* 213. *Fl. Dan. icon.* 487.

RADIX e plurimis fibrillis capillaceis albis constat, quæ in limum profunde dimittuntur.	ROOT consists of numerous white capillary fibres, which penetrate deep into the mud.
CAULIS sive SCAPUS floriferus, pedalis, simplex, erectus, multiflorus, versus apicem glandulis scabriusculus, ad basin foliis plurimis instructus, unde per aquam longe excurrunt caules plures qui fibrillas dimittunt.	STALK or flowering SCAPUS, a foot high, simple, upright, sustaining many flowers, towards the top roughish with little glands, furnished at bottom with numerous leaves, from whence several stalks proceed and run out to a considerable length through the water throwing out numerous white fibres.
FOLIA plurima, plerumque immersa, pinnata, in apicibus caulium juniorum densa, reflexa, Pinnis linearibus planis.	LEAVES numerous, generally under the water, pinnated, growing in tufts on the tops of the young stalks, bending downwards, the Pinnæ linear and flat.
FLORES pallide purpurei, verticillati, spicati, Pedunculi ad. 10, Bracteâ, ad basin instructi, post florescentiam reflexi.	FLOWERS of a pale purple colour, growing in whirls, and forming a spike. Peduncles to 10 in number, furnished at bottom with a Bractæa, when the flowers are gone off turning downwards.
CALYX: PERIANTHIUM monophyllum, quinquepartitum: LACINIIS linearibus, erecto-patulis, *fig.* 1.	CALYX: a PERIANTHIUM of one leaf, divided into five SEGMENTS, which are linear, upright and somewhat spreading, *fig.* 1.
COROLLA: monopetala, hypocrateriformis, TUBUS longitudine calycis, LIMBUS quinquefidus, planus: LACINIIS ovato-oblongis, emarginatis, *fig.* 2.	COROLLA: monopetalous and salver-shaped, the TUBE the length of the calyx; the LIMB divided into five segments and flat; the SEGMENTS of an oval oblong shape with a notch at the extremity, *fig.* 2.
STAMINA: FILAMENTA quinque, subulata, brevia, erecta. ANTHERÆ oblongæ, flavæ. *fig.* 3.	STAMINA: five FILAMENTS tapering, short, and upright, ANTHERÆ oblong and yellow, *fig.* 3.
PISTILLUM: GERMEN subglobosum. STYLUS filiformis, brevis. STIGMA globosum, *fig.* 4.	PISTILLUM: GERMEN roundish, STYLE thread-shaped and short, STIGMA spherical, *fig.* 4.
PERICARPIUM: CAPSULA globosa, unilocularis, subpellucida, *fig.* 5.	SEED-VESSEL: a round CAPSULE of one cavity, slightly transparent, *fig.* 5.
SEMINA plurima, ovata, pallide fusca, *fig.* 7. receptaculo globoso intra capsulam affixa, *fig.* 6.	SEEDS numerous, oval, of a pale brown colour, *fig.* 7: affixed to a round receptacle within the capsule, *fig.* 6.

This singular plant abounds in most of our watry Ditches, particularly in such as divide the Meadows, and flowers in May and June, continuing for a considerable time in blossom; among a variety of other places it may be found in a ditch on the right hand side of the Field Way leading from Kent-street Road to Peckham.

We do not find any author that mentions its possessing any properties to recommend it but its beauty and singularity, both of which it possesses in a degree sufficient to command our admiration.

The leaves generally grow beneath the surface of the water and afford a Nidus if not Nourishment to the fresh-water Periwinkle and some other small shell fish.

Antient Botanists have given it the names of *Millefolium aquaticum*, and *Viola aquatica*; the great number of its leaves induced them, with some propriety, to call it *Millefolium*, but why they should call it a *Viola* seems difficult to determine, as the blossom has nothing in its structure similar to the flowers of that Genus. Boerhave afterwards called it *Hottonia*, in honour of Dr. Hotton, which name Linnæus has continued.

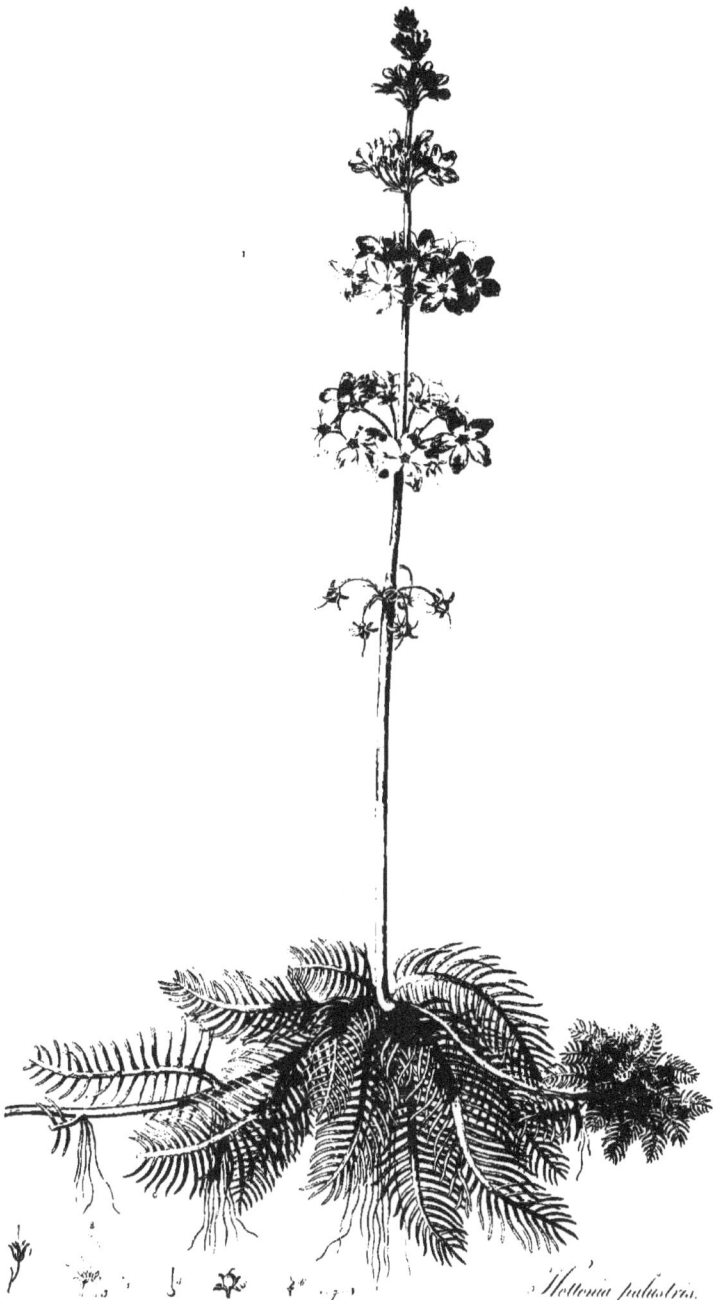

Hottonia palustris.

ANAGALLIS ARVENSIS. PIMPERNEL.

ANAGALLIS *Linnæi Gen. Plant.* PENTANDRIA MONOGYNIA.

Raii Gen. 18. HERBÆ FRUCTU SICCO SINGULARI, FLORE MONOPETALO.

ANAGALLIS foliis indivisis caule procumbente. *Lin. Spec. Plant.* 211.

ANAGALLIS phœniceo flore. *Bauhin. pin.* 252.

ANAGALLIS mas *Fuschii* 18. *Gerard emac.* 617. *Parkinson* 558. *Oeder. Flor. Dan. tab.* 88. *Raii Syn.* 282.
Hudson. 73. *Haller. hist.* 621. 626. *Scopoli. Fl. Carniol.* 139.

RADIX simplex, fibrosa, annua.

ROOT simple, fibrous, annual.

CAULIS ramosus, prostratus, quadrangularis, lævis, subtortuosus, *fig.* 1.

STALK branched, procumbent, quadrangular, smooth, and a little twisted, *fig.* 1.

FOLIA opposita, sessilia, cordata, glabra, *subtus punctis fuscis notata.*

LEAVES opposite, sessile, heart-shaped, smooth, *underneath dotted with brown.*

PEDUNCULI oppositi, foliis fere duplo longiores, inflexi.

PEDUNCLES opposite, nearly twice the length of the leaves, bending downwards.

CALYX persistens, quinquepartitus, segmentis triangularibus, alatis, membranaceis, *fig.* 2.

CALYX persisting, divided into five Segments, the Segments triangular and membranous at the edges, *fig.* 2.

COROLLA monopetala, quinquepartita, laciniis rotundis, coccineis, ad basin purpureis, margine crenatis, subpilosis, *fig.* 3. 4.

COROLLA monopetalous, quinquepartite, the laciniæ scarlet purplish at bottom, the edges slightly notched, and hairy, *fig.* 3. 4.

STAMINA. FILAMENTA quinque, erecta, *pilosissima, (pili articulati!)* superne purpurea. ANTHERÆ oblongæ, biloculares, flavæ, insidentes, *fig.* 5. 6.

STAMINA. five FILAMENTS upright and *very hairy (the hairs when magnified jointed!)* at top purplish: The ANTHERÆ oblong, bilocular, yellow, and sitting on the filaments, *fig.* 5. 6.

PISTILLUM. GERMEN rotundum, STYLUS filiformi:, *obliquus,* longitudine filamentorum; STIGMA subrotundum, *extra circulum staminum locatum,* *fig.* 7.

PISTILLUM. the GERMEN round, the STYLE filiform, the length of the filaments, the STIGMA roundish, *placed without the circle of the Stamina, fig.* 7.

PERICARPIUM. CAPSULA rotunda, nitida, quinquenervia, subdiaphana, circumscissa, fusca, *fig.* 8.

SEED-VESSEL. a CAPSULE, round, shining, brown, slightly transparent, having five nerves, dividing into two equal parts, *fig.* 8.

SEMINA plurima, angulosa, fusca, *fig.* 9.

SEEDS numerous, brown, and angular, *fig.* 9.

NATURE seems to have taken uncommon pains in the formation of the Flowers of this little Plant; few possess more liveliness of colour or greater delicacy of structure, this must be sufficiently obvious to every common observer, but when its minute parts come to be viewed by the microscope, we are charmed with beauties altogether novel and unexpected; we then find that the edges of the flowers, which to the naked eye appear a little uneven or hairy, are furnished with a number of little glands placed on footstalks, and that the hairs of the Filaments which partly tend to distinguish this Genus, are regularly jointed: the Pistillum which generally arises upright betwixt the Stamina, is here inclined to one side, so that the Stigma is placed without the circle of the Stamina: The care which Nature has taken likewise in the preservation of these delicate parts from the injury of the weather, is not less remarkable. Every morning, if the weather be fair and warm, the blossoms fully expand, but if rain falls, or there be much moisture in the air, the flowers quickly close themselves up to secure the inclosed Antheræ and Stigma, from having their functions destroyed: from this property which it has in common with many plants of the same class, it has acquired the name of the *Shepherd's,* or *Poor Man's Weather-glass,*—they have remarked that if the flowers be open in a morning it will prove a fine day, if shut, the contrary.

The small birds (PASSERES LINNÆI) are fond of the seeds of this Plant; and according to experiments made by some of LINNÆUS's pupils, it appears that kine and goats feed on it.

It is very common in corn fields and gardens, flowering all the Summer.

A variety with four leaves at a joint sometimes occurs in rich soil; but as it differs in no other part, and is a mere variety, it scarcely deserves a distinct figure. It is also found with blue and sometimes with white flowers, but we have not observed either of these varieties near London.

Anagallis arvensis.

Convolvulus Sepium. Large white Convolvulus

or great Bindweed.

CONVOLVULUS *Linnæi. Gen. Pl.* Pentandria Monogynia.

Raii Syn. Gen. 18. Herbæ fructu sicco singulari flore monopetalo.

CONVOLVULUS *(sepium)* foliis sagittatis, postice truncatis, pedunculis tetragonis, unifloris. *Linn. Syst.*

Vegetab. p. 168. *Fl. Suecic. p.* 64.

CONVOLVULUS foliis sagittatis, hamis emarginatis, angulosis, petiolis unifloris, stipulis cordatis maximis.

Haller. hist. V. 1. *p.* 295.

CONVOLVULUS Major albus. *Bauhin. pin.* 294.

SMILAX lævis major. *Gerard emac.* 861. *Parkinson.* 163. *Raii Syn. p.* 275. Great Bindweed. *Hudson.*

Fl. Angl. p. 74. *Scopoli. Fl. Carniol.* 141. *Fl. Dan. icon.* 458.

RADIX perennis, crassitie pennæ anserinæ, alba, sub terra reptans et late se propagans, vix eradicanda, Hortorum pestis.

ROOT perennial, about the thickness of a goose quill, of a white colour, creeping under the ground and propagating itself exceedingly, rooted out with the greatest difficulty, and hence very troublesome in Gardens.

CAULES numerosi, volubiles, tortuosi, striati, orgyales, subramosi.

STALKS numerous, twining, twisted, striated, generally about six feet high and somewhat branched.

RAMI pauci, alterni, cauli similes.

BRANCHES few, alternate, like the Stalk.

FOLIA alterna, sagittata, postice truncata, glabra, petiolata.

LEAVES alternate, arrow-shaped, apparently cut off behind, smooth, and placed on foot-stalks.

PEDUNCULI uniflori, alterni, tetragoni.

FOOT-STALKS of the flowers, alternate, supporting one flower only, and four square.

CALYX Involucrum biphyllum, foliolis oblongo-cordatis, subcarinatis, venosis, purpurascentibus. *fig.* 2.

CALYX an Involucrum composed of two heart-shaped leaves, slightly keel-shaped, veiny, and purplish. *fig.* 2.

CALYX Perianthium pentaphyllum, tubulosum, foliolis ovato-lanceolatis, pallide virentibus. *fig.* 1.

CALYX a Perianthium, composed of five leaves and tubular, the leaves of an oval pointed shape and pale green colour. *fig.* 1.

COROLLA monopetala, infundibuliformis, lactea, limbo lato, obscure diviso, paululum reflexo.

COROLLA monopetalous, funnel shaped, of a white colour, the limb broad, obscurely divided, and turned back a little.

STAMINA: Filamenta quinque, fundo corollæ inserta, hirsutula, alba, subulata; Antheræ sagittatæ, albæ, insidentes. *fig.* 3.

STAMINA: five Filaments inserted into the bottom of the corolla, slightly hairy, white and tapering, the Antheræ arrow shaped, white, and fitting on the filaments. *fig.* 3.

PISTILLUM: Germen subovatum; Stylus subulatus apice tortuosus; Stigma bifidum. *fig.* 4. 5.

PISTILLUM: Germen somewhat oval, Style tapering, twisted at top; the Stigma bifid. *fig.* 4. 5.

NECTARIUM: Glandula crocea annuliformis ad basin Germinis.

NECTARY a yellow gland surrounding the base of the Germen.

PERICARPIUM: Capsula subrotunda, fuliginosa, mucronata *fig.* 6. 7.

SEED-VESSEL a roundish Capsule of a footy colour and pointed. *fig.* 9. 7.

SEMINA angulosa, fusca, Cotyledonibus mire convolutis. *fig.* 8. 9.

SEEDS angular and brown, the Cotyledons folded up in a very singular manner. *fig.* 8. 9.

The plant which produces the *Scammony* is a species of Convolvulus, very similar to that which we have now described, hence Dr. Cullen and some other Physicians have conjectured that our Convolvulus might possess similar properties, but if it should be found to contain such properties, the smallness of it roots would prevent its juice from being collected in the same manner with that which flows on incision from the large root of the Scammony plant, and which hardens and forms that purgative substance. Whether an extract made from the expressed juice of the roots, or any other preparation of them might possess a purgative property, or if it should, whether such a purgative would be so far superior to any now in general use as to introduce it deservedly into practice, is what we cannot pretend to decide on. Hogs are said to eat and even to be fond of the roots.

It grows exceedingly common in our hedges, and flowers in August and September. Where it has once gained ground it is with the greatest difficulty eradicated; was it not for this property and its being so common, it would doubtless be considered, as it really is, a very ornamental plant.

My ingenious Friend Mr. Church, Surgeon, at Illington, (who has taken much pains to collect and acquire a knowledge of our English Insects) informs me that the Caterpillar of the *Phalæna Vibicoria* or *Bloody vein Moth,* (vid. Clerc. Phalen. pl. 3. fig. 1.) feeds on this plant, and the *Sphinx Convolvuli* or *Unicorn Hawk Moth,* (vid. Roesel. Cl. 1. pap. noct. t. 7.) is well known to take its name from feeding on this plant also.

Convolvulus Sepium.

Solanum Dulcamara.

SOLANUM DULCAMARA. WOODY NIGHTSHADE.

SOLANUM *Linnæi Gen. Pl.*. PENTANDRIA MONOGYNIA.

Raii Gen. 16. HERBÆ BACCIFERÆ.

SOLANUM *Dulcamara* caule inermi frutefcente flexuofo, foliis fuperioribus haftatis, racemis cymofis. *Lin. Sp. Pl.* 264.

SOLANUM Scandens feu Dulcamara *Bauhin. Pin.* 176. Amara Dulcis. *Gerard. emac.* 350. Solanum lignofum *Parkinfon.* 350. *Raii Synopfis* 265. *Hudfon. Flor. Angl.* p. 78. *Scopoli Flor. Carniol.* p. 161. *Haller. hift. Plant. Helv.* p. 248.

RADIX *perennis.*

CAULIS fruticofus, fcandens, fiftulofus, ramofus, tuberculis parvis fubafper, leniter angulofus, orgyalis et ultra.

RAMI alterni, juniores purpurei.

FOLIA petiolata, mollia, venofa, in caulem fubdecurrentia, *inferiora* ovato-lanceolata, integerrima ; *fuperiora* trilobo-haftata.

FLORES in CYMAS racemofas difpofiti ; pedunculi florales ad bafin bulbofi, aut ex acetabulo quafi prodeuntes.

CALYX. PERIANTHIUM monophyllum, parvum, quinquefidum, purpureum, fegmentis obtufiusculis, perfiftens ; *fig.* 1.

COROLLA monopetala, rotata ; TUBUS breviffimus ; LIMBUS quinquepartitus, LACINIIS lanceolatis, purpureis, reflexis ; FAUX nigra, nitida, ad bafin fingulæ laciniæ maculæ duæ, virides, *fig.* 3—2.

STAMINA. FILAMENTA quinque, breviffima, tubo corollæ inferta, nigro-purpurea. ANTHERÆ quinque, flavæ, erectæ, in tubum fubconicum coalitæ, apicibus biforaminulis, *fig.* 4—5.

PISTILLUM. GERMEN pyriforme ; STYLUS fubulatus, ftaminibus paulo longior ; STIGMA fimplex, obtufum, *fig.* 6.

PERICARPIUM. BACCA ovata, coccinea, glabra, bilocularis, receptaculo utrinque convexo, cui femina adnectuntur, *fig.* 8.

SEMINA plures, lutefcentia, compreffa, fubreniformia, pulpo odoris ingrati obtecta, *fig.* 9.

ROOT *perennial.*

STALK woody, climbing, hollow, branched, thinly befet with fmall pointed tubercles, flightly angular, and growing to the height of fix feet, or more.

BRANCHES alternate, the younger ones purple,

LEAVES ftanding on foot-ftalks, of an oval pointed fhape, foft, veiny, running flightly down the ftalk, the *lower ones* entire, the *upper ones* halbert fhaped.

FLOWERS growing in branched CYMÆ, the proper peduncles of the flowers bulbous at their bafe, or growing out of a kind of focket.

CALYX a PERIANTHIUM of one leaf, fmall, and purple, divided into five fegments, the fegments bluntifh, perfifting ; *fig.* 1.

COROLLA monopetalous, wheel-fhaped, the TUBE very fhort, the limb divided into five fegments, the SEGMENTS lancet-fhaped, purple, and turning back ; the MOUTH black and fhining, at the bottom of each fegment are two roundifh green fpots, *fig.* 3. 2.

STAMINA. five FILAMENTS very fhort, of a black purple colour, and inferted into the tube of the Corolla. Five ANTHERÆ yellow, upright, and uniting into a tube, with two holes at the top of each, out of which the POLLEN is difcharged, *fig.* 4. 5.

PISTILLUM. the GERMEN pear-fhaped ; the STYLE tapering, a little longer than the Stamina ; the STIGMA fimple and obtufe ; *fig.* 6.

PERICARPIUM. an oval, fcarlet, fmooth BERRY of two cavities, the receptacle to which the feeds are connected, is round, on both fides ; *fig.* 8.

SEEDS feveral, flat, fomewhat kidney-fhaped ; *fig.* 9. of a yellowifh colour, inclofed in the pulp, which has a difagreeable fmell.

THE Woody Nightfhade has been commended as a medicine for many diftempers by the old Botanifts, in their ufually lavifh manner ; but PARKINSON fays, he found the juice of it prove a very churlifh purge. LINNÆUS prefers an infufion of the ftalk of this plant to any of the foreign woods, as a cleanfer of the blood, and recommends it in inflammatory fevers, obftructions, the itch, and rheumatifm ; and to render the knowledge of plants as extenfively ufeful as poffible, he does not think it beneath him to remark, that the *Swedifh* Peafants make hoops of the ftalk of this plant to bind their wooden cans. RAY informs us, that the inhabitants of, *Weftphalia,* who are fubject to the fcurvy, make ufe of a decoction of the whole plant as their common drink with fuccefs againft that diftemper.

FLOYER fays, that thirty berries of this plant killed a dog in lefs than three hours, and remained undigefted in his ftomach ; as thefe berries from their refemblance, may happen by miftake to be eaten for currants by children, it may not be improper to remark, that in fuch a cafe, it is advifeable to pour down inftantly as much warm water as poffible, to dilute the poifonous juice, and provoke vomiting, till farther affiftance can be had.

Goats and fheep are faid to feed on this plant, but our other cattle, *viz.* kine, horfes, and fwine, refufe it.

It grows plentifully in moift hedges, and blows from *July* to *Auguft.* The berries are ripe in *September* and *October.* It is fometimes found with a white flower.

LONICERA *PERICLYMENUM.* HONEY SUCKLE

or WOODBINE.

LONICERA *Linnæi Gen. Pl.* PENTANDRIA MONOGYNIA.

Raii Synopfis. ARBORES ET FRUTICES FRUCTU FLORI PETALOIDI CONTIGUO.

LONICERA capitulis ovatis imbricatis terminalibus, foliis omnibus diftinctis. *Lin. Sp. Pl.* 247.

PERICLYMENUS *Fufchii Icon* 646.

PERICLYMENUM non perfoliatum Germanicum. *Bauhin pin.* 302.

CAPRIFOLIUM Germanicum *Dodon. Gerard. emac.* 891. *Parkinfon.* 1460. *Raii Synop.* 458. *Hudfon Fl.* 80.

Haller. hift. 301. *Scopoli Fl. Carniol. p.* 153.

CAULIS lignofus, volubilis, orgyalis et ultra ; cortice pallide fufco; RAMI oppofiti, purpurei.

STALK woody, twining, growing to the height of fix feet or more, the Bark a pale brown, the BRANCHES oppofite and purple.

FOLIA oppofita, ovata, glabra, fubtus cærulescentia.

LEAVES oppofite, oval, fmooth, underneath of a blueifh colour.

FLORES terminales, verticillatim difpofiti, patentes, rubri, interne flavi, odoratiffimi.

FLOWERS terminal, growing in a whirl, and fpreading ; externally red, internally yellow, and fragrant.

CALYX PERIANTHIUM fuperum, breviffimum, quinquepartitum ; fegmentis ovato-lanceolatis, erectis, duobus inferioribus remotioribus. *fig.* 1.

CALYX a *Perianthium* placed above the Germen, very fhort, divided into five fegments, which are of an oval pointed fhape, and upright, the two inferior ones moft remote from each other. *fig.* 1.

BRACTEÆ fubcordatæ, *fig.* 8, germina imbricatim cingentes, ad marginem præcipue fcabræ, ut funt calyx, et tubi bafis pilis glanduliferis.

FLORAL LEAVES laying one over the other, and clofely embracing the Germina, reddifh at the edges and cover'd as well as the Calyx and bafe of the tube with glandular hairs. *fig.* 8.

COROLLA monopetala, tubulofa, TUBUS oblongus, fubinfundibuliformis, LIMBUS bipartitus, laciniis revolutis, fuperiore quadrifida, fegmentis fere æqualibus, obtufis, inferiore integra. *fig.* 2.

COROLLA monopetalous, and tubular, the TUBE long, and fomewhat funnel fhaped, the LIMB bipartite, the laciniæ rolling back, the upper one divided into four blunt and nearly equal fegments, the lower one entire. *fig.* 2.

STAMINA : FILAMENTA quinque filiformia, corollæ longiora, alba, tubo corollæ inferta ; *fig.* 3. ANTHERÆ dum pollinem involvunt oblongæ, incumbentes, poftea lunatæ. *fig.* 4.

STAMINA : five white FILAMENTS, of an equal thicknefs throughout, longer than the Corolla and inferted into its tube, *fig.* 3. the ANTHERÆ while they contain the Pollen oblong, afterwards femilunar and of a yellow colour. *fig.* 4.

PISTILLUM : GERMEN fubrotundum, inferum, *fig.* 5. STYLUS filiformis, Staminibus paulo longior, *fig.* 6. STIGMA capitatum, fubrotundum, trifidum, viride. *fig.* 7.

PISTILLUM: the GERMEN roundifh and placed below the Calyx, *fig.* 5. the STYLE filiform, a little longer than the Stamina, *fig.* 6. the STIGMA roundifh, trifid, and of a green colour. *fig.* 7.

PERICARPIA : BACCÆ plures, fubrotundæ, rubræ, umbilicatæ, biloculares, omnes diftinctæ. *fig.* 9.

SEED VESSELS feveral roundifh red BERRIES, having the remains of the Calyx adhering to them, and all diftinct. *fig.* 9.

SEMINA Plura, lutefcentia, hinc convexa inde plana. *fig.* 10.

SEEDS feveral, of a yellowifh brown colour, round on one fide and flattifh on the other. *fig.* 10.

The early writers attributed virtues to this officinal plant which the latter have been inclined to give up, as a medicine we muft not expect much from it, but the beauty, fingularity, and exquifite fragrance of its flowers have long given it a place in our Gardens. It is a Climber and turns from Eaft to Weft with moft of our other Englifh Climbers, and in common with them it bears clipping and pruning well ; for in a ftate of nature thofe plants that cannot afcend without twining round others are often liable to lofe large branches, they have therefore a proportional vigour of growth to reftore accidental damages. This plant is fubject when placed near Buildings to be disfigured and injured by fmall infects called Aphides or vulgarly blights, thefe Animalculæ were formerly fuppofed to be brought by the Eaft Wind, and confequently the mifchief was looked upon as inevitable, but obfervation has of late years corected that Error, their hiftory is well known*, but no effectual remedy againft them is as yet difcover'd. Thefe Infects are not very numerous in Spring, but as the Summer advances they encreafe in a furprifing degree ; to preferve the plant therefore from injury it is neceffary to watch their firft attacks, cut off and deftroy the branches they firft appear on, for when they have once gained ground they are defended by their numbers. We have feen fmall plants cleared of them by fprinkling fpanifh fnuff on the infected branches, but for large trees this remedy is fcarcely practicable. The leaves are likewife liable to be curled up by a fmall caterpillar (Phalæna Tortrix *Linnæi*) which produces a beautifull little Moth, fee Albins hiftory of englifh Infects pl. 73. It is fed on by Kine, Goats, and Sheep, but Horfes refufe it.

To fhew the confufion of antient names it may not be improper to mention that this plant and Woodrofie (Afperula odorata) have been both called Matrifylva by the old botanic writers. Our Poets alfo have ftrangely confounded the names of this plant, SHAKESPEAR fays
> So doth the WOODBINE the fweet HONEYSUCKLE
> Gently entwift

MILTON feems to call this plant *Eglantine* although that is an undoubted name for the *Sweet Briar*
> Through the SWEET BRIAR or the Vine
> Or the TWISTED EGLANTINE.

We find it plentifully in Woods and Hedges flowering from July to September, fuch plants as grow in fhady places produce bloffoms of a paler colour, and they univerfally fmell fweeteft in the Evening ; at which time fome particular Species of Sphinges (*Linnæi*) or Hawk Moths are frequently obferved in Gardens hovering over the bloffoms and with their long tongues which are peculiarly adapted to the purpofe, extracting honey from the very bottom of the flowers.

* Vid Reaumur and Geoffroy.

Lonicera Periclymenum.

Hedera Helix.

1 2 3 4 5 6

HEDERA HELIX. IVY.

HEDERA *Limæi Gen. Pl.* PENTANDRIA MONOGYNIA. *Petala* quinque oblonga. *Bacca* quinquesperma calyce cincta.
    Raii Syn. ARBORES ET FRUTICES FRUCTU FLORI PETALOIDI CONTIGUO.
HEDERA *Helix* foliis ovatis lobatifque. *Linn. Syst. Vegetab. p.* 202. *Sp. Pl.* 292. *Fl. Suecic. p.* 75.
HEDERA foliis fterilibus trilobatis, fructiferis ovato-lanceolatis. *Haller hift. helv. n.* 826.
HEDERA *Helix. Scopoli Fl. Carniol. n.* 271. *Hudfon Fl. Angl p.* 85.
HEDERA arborea. *Bauhin. Pin.* 305.
HEDERA poetica. *Bauhin. Pin.* 305.
HEDERA major fterilis. *Bauhin. Pin.* 305.
HEDERA humi repens. *Bauhin. Pin.* 305.
HEDERA arborea five fcandens et corymbofa communis. *Parkinfon* 678.
HEDERA *Helix Ger. Em.* 858. *Raii Syn.* 459. Climbing or Berried Ivy: alfo Barren or Creeping Ivy.

TRUNCUS in arboribus hujus fpeciei fenefcentibus cortice rimofo cinereo veftitur, in novellis ramis viridis aut purpureus cernitur, fibrillas e latere interiori exerit, quorum ope proximis arboribus aut parietibus innixus alta petit.

FOLIA quam maxime varia, dum planta repit plerumque trilobata, quinquelobata etiam occurunt; adminiculis derelictis, ovata fiunt; glabra, nitentia, nunc rubedine ornata, nunc venis albis picta, prefertim in ramulis junioribus.

FLORES lutefcentes, in fummitatibus caulium umbellatim difpofiti, UMBELLÆ denfæ, globofæ.
COROLLA: quinque, ovata, flavefcentia, patentia.

STAMINA: FILAMENTA quinque longitudine Corollæ; ANTHERÆ bafi bifidæ, incumbentes, *fig.* 1.

PISTILLUM: GERMEN turbinatum; STYLUS fimplex, breviffimus; STIGMA fimplex, *fig.* 2.
PERICARPIUM: BACCA globofa, nigra, intus purpurea, quadrilocularis aut quinquelocularis, coronata receptaculo et ftylo conico brevi, loculis monofpermis, *fig.* 3, 4.
SEMINA quinque, hinc gibba, inde angulata, *fig.* 6.

TRUNK; the trunk in trees of this fpecies, which are old, is covered with an afh-coloured chopped bark, in the young branches it is of a green or purple colour; from the infide of the trunk a great number of fmall fibres are thrown out, by the affiftance of which, it fupports itfelf on the neareft walls and trees, and climbs aloft.

LEAVES as various as poffible, while the plant creeps they are in general trilobate, fometimes quinquelobate, leaving its fupporters, they become oval; fmooth, fhining, fometimes tictured with red, fometimes painted with white veins, particularly in the young branches.

FLOWERS yellowifh, growing on the top of the ftalks in thick round UMBELS,
COROLLA: PETALS five, oval, yellowifh and fpreading.

STAMINA: five FILAMENTS the length of the Corolla; ANTHERÆ bifid at bottom, and incumbent, *fig.* 1.

PISTILLUM: GERMEN roundifh; STYLE fimple and very fhort; STIGMA fimple, *fig.* 2.
SEED-VESSEL: a round BERRY, externally black, internally purple, with four or five cavities each containing one feed, crowned with the receptacle and fhort conic Style, *fig.* 3, 4.
SEEDS five, on one fide gibbous, on the other angular, *fig.* 6.

The Hedera Helix begins to blow in funny afpects towards the end of September, and according to fituation bloffoms on through October, and November. This plant is one of the laft blowers, and is much reforted to by bees, and flies of various fpecies, which fwarm on its branches, and feed on its bloffoms, making fuch a humming on funny days as may be diftinguifhed at a confiderable diftance.

The berries encreafe in bulk gradually all through the winter months, and are full formed by February; in April they ripen and turn very black, and are eaten by feveral fpecies of thrufhes, and wild pigeons. Thus does fructification manifeftly obtain in this inftance all through the winter months, as well as in the moffes and lichens.

Sheep are very fond of Ivy, which in hard weather is a warm and wholfome food; and therefore fhepherds in fnowy feafons cut down branches for their flocks to brouze on. CATO directs that in a fcarcity of hay, cattle fhould be foddered with Ivy.

Profeffor KALM, in his travels through the greateft part of N. America, faw but one plant of Ivy, and that was running up the walls of a man's houfe: this fpecimen was probably carried thither by fome European who perhaps was defirous of propagating in that new world a plant that might ftill recall to his mind the pleafing Idea of his native cottage, tufted with the foliage of this beautiful Evergreen.

The anticuts held this plant in great efteem; their Heroes and Poets are defcribed as wearing garlands compofed of it. The fuppofition its preventing intoxication is of very early date: Homer therefore mentions his Bacchus as Ivy-crowned and often defcribes his Heroes drinking out of a Cup made of the wood of Ivy. (κισσυβιον.) CATO tells us that with a cup of this kind we may diftinguifh wine that has been adulterated with water, for the wine will be difcharged and the water remain: to fuch an extravagant affertion has this grave author been probably led by relying on the fuppofed antipathy between the wine and ivy: This cup is ftill ufed in fome parts of the kingdom as a remedy for a trembling hand; but rational practice has not admitted any part of the Hedera into the Materia Medica, Ivy-leaves however are faid to be fuccefsfully applied to painful Corns. When it trails on the ground it branches are fmall and weak; and its leaves are divided into three lobes; but when it climbs walls or trees it grows much ftronger and the leaf changes to an oval form: thefe different appearances induced old Botanifts to fuppofe there were two or three different fpecies. In its variegated ftate it fometimes appears almoft white, and may perhaps be the *Hedera alba*, and *pallentes Hederæ* of VIRGIL.

Few people are acquainted with the beauty of Ivy when fuffered to run up a ftake, and at length to form itfelf into a ftandard, the fingular complication of its branches and the vivid hue of its leaves give it one of the firft places amongft evergreens in a fhrubbery; In woods when fuffered to grow large and rampant, this plant by twining round their bodies does great damage to timber trees; and therefore fhould be carefully deftroyed: but in ornamented Out-lets, where evergreens do not abound, a few trees covered with Ivy have a very pleafing effect, and moreover induce birds of fong to haunt thofe thickets for the fake of the berries and fhelter.

In the Stump of Ivy many birds build their Nefts particularly the Black-bird.

When Ivy is prejudicial, it may eafily be deftroyed, tho' it has fpread to a great height, by cutting through its Trunk, and this fhows that the fibres which the Stalk throws out in fo fingular a manner ferve more to fupport than nourifh it.

The foft wood of Ivy is made ufe of by Shoemakers to give a fmooth edge to their cutting knives.

Conium maculatum.

CONIUM MACULATUM. HEMLOCK.

CONIUM *Linnæi Gen. Pl.* PENTANDRIA DIGYNIA.

Raii. Syn. Gen. 11. UMBELLIFERÆ HERBÆ.

CONIUM *maculatum feminibus ftriatis. Linn. Syft. Vegetab. p.* 229.

CICUTA *Haller. hift. helv. n.* 766. *v.* 1. *p.* 337.

CONIUM *maculatum. Scopoli Fl. Carniol. p.* 207:

CICUTA *major Bauhin. Pin.* 160.

CICUTA *Gerard emac.* 1061.

CICUTA *vulgaris major Parkinfon* 933. *Raii Syn. p.* 215. *Hudfon Fl. Angl. p.* 100. *Storck. Cicut. Suppl. p.* 7. *t.* 1.

RADIX biennis, craffitudine digiti, longa ufque ad pedalem, in crura fæpe divifa, juniori Paftinacæ haud diffimilis, odoris gravis, et faporis fubdulcis: fecundo anno in caulefcente planta fucco fere caret, firma folidiorque evadit.

ROOT biennial, the thicknefs of ones finger, from fix inches to a foot in length, frequently forked, and not unlike that of a young Parfnep, of a difagreeable fmell and fweetifh tafte: in the fecond year of its growth when the plant has a flowering ftem, it becomes drier, more firm and folid.

CAULIS orgyalis, teres, nitidus, lævis, fiftulofus, ad bafin craffitie pollicis, rore glauco tectus, et maculis fanguineis pictus, verfus fummitatem ramofus, et ftriatus.

STALK about fix inches high, round, fhining, fmooth and hollow, at bottom the thicknefs of ones thumb, covered with a bluefh kind of powder which eafily wipes off, and fpotted with red, towards the top branched and ftriated.

FOLIA inferiora magna, etiam bipedalia, atro-virentia, nitentia, multiplicato-pinnata, pinnulis oblongis incifo-ferratis; SPATHA fulcata.

LEAVES. The bottom leaves large, even two feet long, of a dark green colour and fhining, many times pinnated, the pinnulæ oblong and fharply cut in; the SPATHA grooved.

INFLORESCENTIA. *Umbella univerfalis* Radiis plurimis patentibus ftriatis; *partialis* confimilis.

INFLORESCENCE. The *Univerfal Umbell* is compofed of many ftriated and fpreading Radii; the *Partial Umbell* fimilar to it.

CALYX: *Involucrum univerfale* e foliolis 5—7 conftat, lanceolato acuminatis, reflexis, margine albidis, *fig.* 1; *partiale* 3 aut 4 dimidiatis, extrorfum patentibus, *fig.* 2.

CALYX: the *Univerfal Involucrum* confifts of 5 or 7 leaves, which are lanceolate, turned back, and whitifh at the edges, *fig.* 1: the *Partial Involucrum* is compofed of 3 or 4 leaves, which furround one half of the ftalk only, and fpread outward, *fig.* 2.

COROLLA: PETALA quinque, alba, inæqualia, inflexo-cordata, *fig.* 3.

COROLLA: PETALS five, white, unequal, heart-fhaped, and bent in at top, *fig.* 3

STAMINA: FILAMENTA quinque, alba, longitudine Corollæ; ANTHERÆ albæ, *fig.* 3.

STAMINA: FILAMENTS five, white, the length of the Corolla; ANTHERÆ white, *fig.* 3.

PISTILLUM: STYLI duo, albi, filiformes, non admodum breves; STIGMATA fubrotunda; GERMEN inferum, ftriatum, *fig.* 3, 4.

PISTILLUM: GERMEN beneath the Corolla, ftriated, *fig.* 3, 4; STYLES two, filiform and not very fhort; STIGMATA round, *fig.* 3.

FRUCTUS fubrotundus, e binis feminibus fufcefcentibus componitur, hinc planiufculis, illinc gibbis, cum *ftriis* quinque elevatis crenulatis, *fig.* 4, 5.

FRUIT is roundifh, and compofed of two brownifh feeds, flattifh on one fide and round on the other, with five notched and elevated ridges, *fig.* 4, 5.

THE powerfull deleterious properties of this herb have been long known, and acknowledged by all botanic writers; whence it has been commonly ranged in the clafs of Vegetable Poifons. And as fuch active principles under fkilful management, are likely to afford the moft efficacious remedies, this plant has been alfo admitted as an article of the Materia Medica. Until lately however, the ufe of it was chiefly confined to external applications, where its narcotic qualities may undoubtedly affift in affwaging pain, forwarding fuppuration, &c. But in the year 1760, Dr. STORCK, a famous Practitioner at Vienna, publifhed a treatife on Hemlock, recommending an extract made of the infpiffated juice of the herb to be taken internally, from four grains to fixty, or upwards, every day, as a cure for the *Schrophula, Cancer,* and others of the moft terrible and inveterate diforders incident to the human body.

Our Phyficians though laudably cautious of admitting or trufting to novelties, received Dr. STORCK's publication with uncommon ardour, and perhaps no new medicine was ever more immediately or generally tried than this *Extractum Cicutæ.* The fuccefs however not anfwering their expectation, led fome to think they had miftaken the plant. The Author was applied to, and this produced a fupplement (printed 1764) wherein the fpecies is figured, and clearly fhewn to be the *Conium maculatum* of LINNÆUS. It were to be wifhed this had cleared up all difficulties. In his firft treatife the Doctor tells us that the frefh root fliced, yielded a bitter acrid milk, of which a fingle drop or two being applied to the tip of his tongue, prefently rendered it painful, rigid, and fo much fwelled that he could not fpeak. Yet it is certain that the roots of our *Hemlock* may be chewed and fwallowed in confiderable quantities without producing any fenfible effect. Mr. ALCHORNE (who I believe was the firft that laudably exerted himfelf in invefigating this matter,) affures me that he has tried this in every feafon of the Year, and in moft parts of our Ifland, without finding any material difference: and that

he

he has also been well informed both from *Berlin* and *Vienna*, that the *Hemlock* Roots in those countries, are no more virulent than ours about *London*. Mr. TIMOTHY LANE informs me, that he also with great caution made some experiments of the like kind, and in a short time found he could venture to eat a considerable part of a root without any inconvenience; after that, he had some large roots boiled, and found them as agreeable eating at dinner with meat, as Carrots, which they in taste somewhat resembled: and as far as his experience, joined with that of others informed him, the Roots might be cultivated in Gardens, and either eaten raw like Celery, or boiled as Parsnaps or Carrots. That in Spring and Winter they are not woody as in Summer: that he has eaten them from different places and in all seasons; and that he perceived some roots were more pungent than others, but not in any degree worthy notice.

The experiments of these ingenious Gentlemen sufficiently evince the innocence of the rooots of this plant, contrary to what has been asserted by Dr. STÖRCK, and hence we may infer that whatever accounts have been related by Authors of their poisonous qualities, the Roots of some other Plant must have been made use of. In the poisonous quality of the *Herb* however all Authors seem agreed, but with respect to its efficacy as a medicine they very much differ. If we may believe Dr. STÖRCK, there is scarce a disease incident to the human body which it either does not cure, or relieve; but it is remarkable that a copious experience of fifteen years, as well in the great Hospitals of this Metropolis as in the private practice of the whole Kingdom, should not have afforded one instance of a perfect cure by the Extract, at least none such has appeared among the valuable collections of cases published by our College of Physicians and other Medical Societies. Both Dr. FOTHERGILL of London, and the late Dr. RUTTY of Ireland, men of the greatest eminence in their profession, have declared that the success attending it has not been equal to what they had reason to expect from Dr. STORCK's account of it ; *(vid. medical observations and enquiries, vol. 3.)* yet tho' it had failed them in the cure of many of those diseases which unfortunately were the *opprobia medicorum*, it had proved beneficial in various obstinate complaints ; Scrophulous tumours were to appearance dissolved by it; the progress both of occult and ulcerated Cancers was retarded, the pain alleviated, and the discharge changed for the better in every respect; divers putrid and sordid Ulcers were by the use of *Hemlock* remarkably mended in their discharge, and disposed to heal, in some of which the Sublimate had been given in vain; hence the Extract is still frequently used, and will probably continue to be preferred, because its effects as an Anodyne will often afford at least a temporary relief, and because in desperate diseases a doubtful remedy seems better than none at all.

The taking of the Extract is generally attended with a giddiness and often with a pain of the head, nausea, and other disagreeable symptoms; in some however its effects are apparently anodyne, as it eases pain and promotes rest even where Opium has failed.

Physicians seem somewhat divided about the best mode of exhibiting this medicine, some recommending the extract as being most easily taken in the form of pills, others the powder, as not being subject to that variation which the extract is liable to from being made in different ways. With respect to the period likewise at which the plant should be gathered, they seem not perfectly agreed, some recommending it when in its full vigour, and just coming into bloom, others when the flowers are going off and the whole plant has acquired a yellowish hue. That the Extract might be at all times equally active, and uniformly prepared, Dr. CULLEN has for many years recommended the making it from theunripe seeds, and this mode the College of Physicians at Edinburgh has thought proper to adopt in their new Pharmacopœia.

Hemlock grows very frequently on banks by the sides of Roads, by hedge sides, and in Fields and Gardens, flowering in the month of July.

We have a common English Proverb that *what is one Mans Meat is another mans Poison*, and agreeable to this are the lines of LUCRETIUS which relate to this plant ;

"Pinguescere sæpe Cicuta
"Barbigeros pecudes homini quæ est acre venenum."

That it affords nourishment to Birds likewise there is sufficient evidence, our learned Philosopher and accurate Naturalist Mr. RAY, found in the Crop of a Thrush abundance of *Hemlock* seeds, at a time too when other vegetable food might be had in abundance. It appears to be eaten by very few or no Insects.

The dried stems or kexes are used by Boys for various purposes.

The *Hemlock* is obviously distinguished from our other umbelliferous plants by its *large* and *spotted stalk*, by the *dark* and *shining green colour of its bottom leaves*, and particularly by their *disagreeable smell* when bruised, and which according to Dr. STORCK resembles that of Mice. The *Fools Parsley* and *Scandix with rough seeds* are the most likely to be mistaken for this poisonous plant, but may easily be distinguished if attention be paid to the descriptions and figures we have already given of them,

Æthusa Cynapium.

ÆTHUSA *CYNAPIUM*. FOOL's PARSLEY.

ÆTHUSA *Linnæi Gen. Pl.* PENTANDRIA DIGYNIA.
    Raii Syn. Gen. 11. UMBELLIFERÆ HERBÆ.
ÆTHUSA (*Cynapium*) foliis conformibus. *Linnæi Syst. Vegetab. p.* 236. *Flor. Suecie. p.* 98.
ÆTHUSA *Haller. hist. n.* 765.
CICUTA minor petroselino similis. *Baubin. Pin. p.* 160.
CICUTARIA Apii folio. *I. Baubin.*
CICUTARIA tenuifolia *Gerard. emac.* 1063.
CICUTA minor sive fatua *Parkinson.* 933. *Raii Syn. p.* 215. the lesser Hemlock or Fool's Parsley. *Scopoli Fl. Carniol. p.* 206. *Hudson Fl. Angl. p.* 107. *Hill's British Herbal* small Hemlock *tab.* 58. *icon pessima.*

RADIX annua, fusiformis, alba, *minimi digiti crassitudine*, paucis fibris instructa.

ROOT *annual*, tapering, of a white colour, *about the thickness of the little finger*, furnished with few fibres.

CAULIS pedalis ad bipedalem, erectus, ramosus, striatus, fistulosus, glaucus, versus basin sæpe purpureus, *non vere maculatus.*

STALK from one to two feet high, upright, branched, striated or slightly grooved, hollow, covered with a blueish kind of powder which easily wipes off, towards the bottom frequently of a purple colour, *but not spotted.*

FOLIA radicalia et ramea conformia, lævia, superne *atro-virentia*, inferne pallidiora, nitentia, duplicato-pinnata, pinnis pinnatifidis, profunde incisis, pinnulis ovato-acutis, mucronatis. *Vaginæ* ad basin petiolorum parvæ, læves, marginibus membranaceis.

LEAVES: the bottom leaves and those of the branches similar, smooth, on the upper side of a *dark green colour*, underneath paler and shining, twice pinnated, the leaves pinnatifid and deeply cut in, the small leaves or pinnulæ oval and terminating in a fine point. The SHEATHS at the base of the foot-stalks small, smooth and membranous at the edges.

PETIOLI erecti, sulcati.

FOOT-STALKS of the flowers, upright and grooved.

UMBELLA *universalis* patens, radiis interioribus per gradus brevioribus, intimis brevissimis; *partialis* universali similis.

UMBEL: the *universal* umbel spreading, the inner radii gradually shorter, the inmost very short; the partial umbel like the universal.

INVOLUCRUM *universale* nullum, *partiale* dimidiatum, extus positum, *foliolis tribus longissimis linearibus pendulis, fig.* 1.

INVOLUCRUM: the *universal* INVOLUCRUM wanting, the *partial* one placed externally, and only surrounding one half of the umbel, composed of *three very long, linear, and pendulous leaves, fig.* 1.

COROLLA: PETALA quinque, alba, obcordata, inæqualia, apice inflexa, exteriora majora, *fig.* 2.

COROLLA: five unequal, heart-shaped, white PETALS, bent at top, the outer ones largest, *fig.* 2.

STAMINA: FILAMENTA quinque, alba, longitudine corollæ, inflexa: ANTHERÆ albæ, nonnunquam rubellæ, *fig.* 3.

STAMINA: five white FILAMENTS the length of the Corolla, bending in: ANTHERÆ white, sometimes reddish, *fig.* 3.

PISTILLUM: GERMEN inferum, glandulâ virescente coronatum: STYLI duo, primum erecti, dein deflexi: STIGMATA obtusa, *fig.* 4.

PISTILLUM: GERMEN placed below the corolla, and crowned by a glandular substance of a greenish colour: two STYLES first upright, afterwards bending downward: STIGMATA blunt, *fig.* 4.

PERICARPIUM nullum: FRUCTUS ovato-subrotundus, striatus, bipartibilis, *fig.* 5.

SEED-VESSEL wanting: the FRUIT or unripe seed of an oval roundish shape, striated, and dividing into two parts, *fig.* 5.

SEMINA duo, pallide fusca, hinc convexa, profunde striata, hinc plana, figurâ ovato-acutâ notata, *fig.* 6.

SEEDS two, of a pale brown colour, convex and deeply striated on one side, flat on the other, and marked with a figure of an oval pointed shape, *fig.* 6.

ONE of the principal advantages resulting to mankind from Botany, is the rightly ascertaining those plants which are used for food, from those which are known to be poisonous. It not unfrequently happens that both these kinds of Herbs grow in the same soil, nay often in the same bed together, and so similar are they in their general appearance, that the indiscriminating eye of the common observer readily mistakes the one for the other, and hence diseases fatal in their consequences sometimes ensue. To point out then the most obvious distinctions between such kinds of plants, is not only our business but our duty.

The *Fool's Parsley* seems generally allowed to be a plant which possesses poisonous qualities.

Baron HALLER has taken a great deal of pains to collect what has been said concerning it, and quotes many authorities to shew that this plant (on being eaten) has been productive of the most violent symptoms, such as anxiety, hickcough, and a delirium even for the space of three months, stupor, vomiting, convulsions and death: He suspects however that the common Hemlock may sometimes have had a share in producing these symptoms, as he finds in authors that the Fool's Parsley had been used by a whole family without any bad effect, although he imagines this might be owing to the smallness of the quantity eaten. As a corroborating proof of its deleterious quality, LINNÆUS asserts that it proves fatal to geese if they happen to eat it.

Altho' it seems rather doubtful whether it be so poisonous to mankind as is represented, yet it will perhaps be most prudent to consider it as such, until future experiments shall determine its effects with more certainty.

The plants to which this bears the greatest resemblance, are *common Garden Parsley* and *common Hemlock, Conium maculatum*; this similarity has been observed by most Botanic Writers, some of whom have called it a kind of Hemlock, others a kind of Parsley; it differs however considerably from both these Genera. The colour of its leaves alone, is nearly sufficient to distinguish it from Parsley; those of common Parsley are of a *yellowish green colour*, those of Fool's Parsley of a *very dark green*, and much more finely divided; the leaves of Parsley when bruised have a *strong but not disagreeable smell*, those of Fool's Parsley have *very little smell* in them. These marks if attended to are sufficient to distinguish the *leaves* of these two plants, and in the state of leaves they are most liable to be taken for one another, as they grow together in Gardens. Where much Parsley is used, the Mistress of the house therefore would do well to examine the Herbs previous to their being made use of; but the best precaution will be always to sow that variety called curled Parsley, which cannot be mistaken for this or any other plant.

It is distinguished from *Hemlock* by being in every respect smaller, and not having that strong disagreeable smell which characterizes the leaves of that plant; the stalk likewise is not spotted as in the Hemlock; and lastly it is distinguished from *all our umbelliferous* plants by the *three long, narrow, pendulous leaves* which compose its partial *Involucrum*, and which are placed at the bottom of each of the small Umbels.

It grows very common in Gardens, and all kinds of cultivated ground, and *flowers* in July and August.

Scandix Anthriscus

Scandix Anthriscus. Scandix with rough Seeds.

SCANDIX *Linnæi Gen. Pl.* Pentandria Digynia

Raii Syn. Gen. 11. umbelliferæ herbæ

SCANDIX *Anthriscus* feminibus ovatis hifpidis, corollis uniformibus, caule lævi. *Linnæi Syft. Vegetab. p.* 237. *Flor. Suecic. p.* 93.

CAUCALIS vaginis lanuginofis, foliis triplicato-pinnatis, feminibus roftratis. *Haller hift. n.* 743.

MYRRHIS fylveftris, feminibus afperis. *Bauhin pin.* 160. *Parkinfon* 935. *Ger. emac.* 1038. *Raii Syn. p.* 220. Small Hemlock-Chervil with rough Seeds. *Hudfon Fl. Angl. p.* 108. *Jacquin Flor. Auftriac Vol.* 2. *p.* 35. *tab.* 154.

RADIX annua, parva, albida, fubinfipida.	ROOT annual, fmall, whitifh, with little tafte.
CAULIS pedalis ad tripedalem, fæpe altior, fuberoctus, teres, fiftulofus, lævis, ad genicula tumidus et fubftriatus, plerumque viridis.	STALK from one to three feet high, frequently taller, nearly upright, round, hollow, fmooth, fwelled and flightly ftriated at the joints, and moft commonly green.
FOLIA. Vaginæ ad bafin foliorum magnæ, marginibus lanuginofis ; Folia mollia, tenera, multiplicate pinnata, hirfutula, ex luteo-virentia.	LEAVES. The fheaths formed by the bafe of the leaves are large and downy at the edges : the leaves foft, tender, many times pinnated, flightly hairy, and of a yellowifh green colour.
INFLORESCENTIA *Umbella*. Umbellæ obliquæ, pedunculatæ : Pedunculus univerfalis Radiis brevior, Radii *univerfales* 3—5. glabri, *partiales* 2—6.	INFLORESCENCE an *Umbell*, the Umbells oblique, ftanding on footftalks, the general or univerfal footftalk fhorter than the Radii ; the *univerfal* Radii from 3 to 5, the *partial* Radii from 2 to 6.
CALYX : *Involucrum univerfale* nullum. *Partiale* plerumque pentaphyllum, foliolis lanceolato-acuminatis, ciliatis, perfiftentibus *fig.* 1 :	CALYX. *The univerfal Involucrum* wanting, the *Partial* one generally compofed of five leaves, which are pointed, hairy at the edges, and continue. *fig.* 1.
COROLLA : Petala quinque, minima, fubæqualia, alba, fubcordata, apicibus inflexis. *fig.* 2:	COROLLA : five Petals very minute, nearly equal, white, fomewhat heart fhaped, the tips bending in, *fig.* 2.
STAMINA : Filamenta quinque, petalis paulo breviora ; Antheræ primum virides, dein fufcæ *fig* 3.	STAMINA : five Filaments, a little fhorter than the Petals ; the Antheræ firft green, afterwards brown, *fig.* 3.
PISTILLUM : Germen oblongum, inferum, fubcompreffum, hirfutum, Styli duo breves. *fig.* 5.	PISTILLUM : the Germen oblong, placed beneath the Corolla, flattifh, and rough, two Styles very fhort *fig.* 5.
SEMINA duo, oblonga, e fufco-nigricantia, hinc fulcato-plana, inde convexa, roftrata, pilis rigidis hamatis undique afpera *fig.* 6.	SEEDS two, oblong, of a dark brown colour, on one fide flat and grooved, on the other convex, running out to a point, and prickly with ftiff hooked hairs, *fig.* 6.

THE great fimilarity in the external appearance of a great number of umbelliferous plants, frequently hath been the caufe of miftakes, which have fometimes proved hurtful to the health of individuals. At the fame time that there is no clafs of plants which, at firft fight, appears to the young Botanift more difficult of inveftigation than this, there is none perhaps which affords more conftant, or more certain marks of generic and fpecific difference. Obvious diftinctions may be drawn from the *Stalk* and *Leaves*: in fome the ftalk is fmooth in others, rough ; and in others, more or lefs deeply channeled ; in fome the leaves are very finely divided ; and in others, but coarfely fo ; but the parts of *Fructification* afford the moft pleafing, and fcientific, diftinguifhing marks. The abfence, or prefence of the general and partial *Involucrum* the number, fhape, and fituation of its leaves, the number of the *Radii* which compofe the umbell, the fize and equality of the *Petals*, and the very different appearances of the *Seeds*, all unite to render a knowledge of thefe plants eafily acquired.

Some of the *Umbelliferi* are ufed in food, and others in medicine ; the greateft care will therefore be neceffary in the drawing and defcription of thefe ; and in this, no one feems to have fucceeded fo well as the celebrated Jacquin. In the firft and fecond volumes of his *Flora Auftriaca*, lately publifhed, and which indeed are a moft valuable addition to the ftock of botanic knowledge, a great number of thefe plants are figured and defcribed.

This plant grows very common on dry banks and in hedges, flowers from the beginning to the end of May, and the feeds are ripe in June. When it becomes luxuriant, as it fometimes will from growing in a moift fituation, it puts on fomewhat the appearance of the common Hemlock, but may eafily be diftinguifhed from that poifonous plant, if attention be paid to the following particulars : The leaves of the Hemlock are perfectly fmooth ; thefe have a flight hairinefs, are more finely divided, and of a paler green ; the ftalk of the Hemlock is fpotted ; this is not ; the Hemlock has a general involucrum, which in this plant is wanting ; the feeds of the Hemlock are fmooth, and thefe are rough ; the Hemlock has a ftrong difagreeable fmell ; this not difagreeable, but more like Chervil, to which in its virtues it fhould feem neareft allied.

Alsine media

ALSINE *Linnæi Gen. Pl.* PENTANDRIA TRIGYNIA.

Cal. 5-phyllus. *Petala* 5-æqualia. *Caps.* 1-locularis, 3-valvis.

Raii Syn. Gen. 24. HERBÆ PENTAPETALÆ VASCULIFERÆ.

ALSINE *media. Linnæi Syst. Vegetab.* p. 246. *Flora Suecic.* p. 37.

ALSINE foliis petiolatis, ovato lanceolatis, petalis bipartitis. *Haller hist. helv.* n. 880.

ALSINE *media. Scopoli Fl. Carniol.* n. 376.

ALSINE *media. Bauhin pin.* p. 250.

ALSINE *media seu minor. Gerard emac.* 611. *Raii Syn.* p. 347, Common Chickweed. *Hudson Fl. Angl.* p. 113. *Oeder Fl. Dan.* 525, 438.

RADIX annua, fibrosa, capillacea.

CAULES plures, tenelli, teretes, subrepentes, ramosi, viticulis geniculati, *unifariam hirsuti*, apicibus sensim incrassatis.

FOLIA ovato-acuta, glabra, leviter ciliata; inferiora petiolata, superiora sessilia, connata.

PETIOLI ad basin latiora, hirsuti.

PEDUNCULI uniflori, axillares, hirsuti, peracta florescentia penduli, demum erecti.

CALYX: PERIANTHIUM pentaphyllum, foliolis lanceolaris, concavis, subcarinatis, marginatis, hirsutis, *Petalis longioribus*, fig. 1.

COROLLA: PETALA quinque, alba, nitida, ad basin fere partita, fig. 3, 4, 5.

STAMINA: FILAMENTA quinque, alba, inter Petala locata, Glandulâ ad basin instructâ; ANTHERÆ subrotundæ, purpurascentes, fig. 5, 6.

PISTILLUM: GERMEN subovatum; STYLI tres filiformes; STIGMATA simplicia, fig. 7.

PERICARPIUM: CAPSULA unilocularis, in valvulas sex dehiscentes, fig. 8.

SEMINA octo ad quindecim, subreniformia, aspera, e fusco-aurantiaca, pedicellis receptaculo connexa, fig. 9, 10, aucti.

ROOT annual, fibrous, capillary.

STALKS numerous, tender, round, striking root here and there, branched, jointed and stringy, *hairy on one side only*, growing thicker towards the top.

LEAVES of a pointed oval shape, smooth, slightly hairy at the edges, the lowermost standing on foot-stalks, the uppermost sessile, connate.

FOOT-STALKS of the leaves broadest at bottom, and hairy.

FOOT-STALKS of the flowers, each sustaining one flower, proceeding from the bosoms of the leaves, hairy, when the flowering is over hanging down, finally becoming upright.

CALYX: a PERIANTHIUM of five leaves, each of which is lanceolate, concave, slightly keel-shaped at bottom, with a margin at the edge, hairy, *and longer than the Petals*, fig. 1.

COROLLA consists of five white shining PETALS, divided nearly to the base, fig. 3, 4, 5.

STAMINA: five white FILAMENTS, placed betwixt the Petals, furnished at bottom with a little Gland; ANTHERÆ roundish, of a purplish colour, fig. 5, 6.

PISTILLUM: GERMEN somewhat oval; STYLES three, filiform; STIGMATA simple, fig. 7.

SEED-VESSEL a CAPSULE of one cavity, splitting into six valves, fig. 8.

SEEDS from eight to fifteen, somewhat kidney-shaped, of a brownish orange colour, with a rough surface, connected to the receptacle by little foot-stalks, fig. 9, 10, magnified.

CHICKWEED being a plant which will grow in almost any situation, is consequently liable to assume many different appearances: when it grows in a rich soil, and shady situation, it will frequently become so large as to resemble the *Cerastium aquaticum*; while at other times, on a dry barren wall, its leaves and stalks will be so minute, as to make the young botanist take it for some species different from the common Chickweed: happily however it affords marks which if attended to, will readily distinguish it from the *Cerastium*, and every other plant: exclusive of its differing from the *Cerastium* in its generic character, its Petals are shorter than the leaves of its Calyx; while in the *Cerastium* they are longer; hence a considerable difference will be observable at first sight in the size of the flowers of these two plants: and from all other plants related to it, it may be distinguished by the singular appearance of its stalk, which is *alternately hairy on one side only*.

The most common number of its Stamina with us is five; yet I have often seen it with less, and sometimes with more; and this inconstancy in the number of its Stamina has been noticed by most botanic writers: GOUAN, in his *Flor. Monspel.* mentions from 3 to 10, with as many Pistilla; this circumstance with respect to the number of its Stamina, unfortunately separates it from other plants with which it appears to have by nature a very near relation: but as five Stamina appear to be its most constant number, LINNÆUS could not have placed it amongst those plants with ten Stamina, without doing violence to his system.

Of annual plants there are few more troublesome; it sows itself plentifully in the summer, and remains green throughout the winter, flowering during the whole time, if the weather be mild: but its chief season for flowering is in the spring. In rich garden mould, where the ground is highly cultivated, and in the fields about town, it does a deal of mischief: by the quickness of its growth and the great number of its shoots, it covers and chaoks many young plants; hence it should be carefully weeded from dunghills.

The seeds are very beautiful, and have the greatest affinity to those of the *Cerastium aquaticum*.

When the flowers first open, the foot-stalks which support them are upright; as the flowers go off they hang down; and when the seeds become ripe, they again become erect.

LINNÆUS has observed that the flowers open from nine in the morning till noon, unless rain falls on the same day, in which case they do not open: from what little observations I have made on this plant, it is not subject to be affected precisely in the same manner here, having seen in the month of March, the blossoms continue rather widely expanded after repeated showers of rain.

It is considered as a wholesome food for Chicken and small Birds, whence, as RAY observes, it has obtained its name: boiled it resembles Spinach so exactly as scarcely to be distinguished from it, and is equally wholesome, being a plant which may be procured almost any where very early in the spring, it may be no bad substitute where Spinach or other greens are not to be had in plenty, and much preferable to Nettle-tops and other plant, which the lower sort of people seek after in the spring with so much avidity. Swine are very fond it, and prefer it to Turnep-tops. It is eaten by many Insects, particularly by the Caterpillar of the *Phalæna Villica* or *Cream spot Tyger Moth*, and other hairy Caterpillars of the Tyger kind.

As a medicine it contains no active principle; but is frequently applied to hot, painful, and inflamatory swellings, either by itself, bruised, or mixed with poultices, with good success.

ERICA TETRALIX. CROSS-LEAVED HEATH.

ERICA *Linnæi Gen. Pl.* OCTANDRIA MONOGYNIA.

Cal. 4-phyllus. *Cor.* 4-fida. *Filamenta* receptaculo inferta. *Antheræ* bifidæ. *Caps.* 4-locularis.

Raii Syn. ARBORES ET FRUTICES.

ERICA *tetralix* foliis quaternis ciliatis, floribus capitatis imbricatis.

ERICA *tetralix*, antheris ariftatis, corollis ovatis, ftylo inclufo, foliis quaternis ciliatis, floribus capitatis.

Linn. Syft. Vegetab. p. 302. Fl. Suecic. n. 337.

ERICA ex rubro nigricans fcoparia. *Bauhin Pin.* 486.

ERICA Brabantica folio Coridis hirfuto quaterno. *I. B.* 1. 358.

ERICA pumila Belgarum Lobelio, fcoparia noftras. *Parkinfon.* 1482.

ERICA major flore purpureo. *Gerard emac.* 1382 *Raii Syn. p.* 471, Low Dutch Heath or Befome Heath. *Hudfon Fl. Angl. p.* 144. *Oeder Fl. Dan. icon.* 81.

CAULES fruticofi, dodrantales aut pedales, ramofi, fufci, fcabriufculi ex relictamentis foliorum.	STALKS fhrubby, about nine or twelve inches high, branched, roughifh from the remains of the leaves which have fallen off.
FOLIA *quaterna*, ovato-linearia, patentia, prope flores cauli adpreffa, marginibus inflexis, *ciliatis*, ciliis glandulà terminatis, fuperficie fuperiore plana, inferiore concava.	LEAVES growing by fours, of an oval-linear fhape, fpreading, near the flowers preffed clofe to the ftalk, the edges turned in and *ciliated or hairy*, each of the hairs terminating in a fmall round globule, the upper furface flat, the inferior furface concave.
FLORES fecundi, imbricati, in capitulum congefti, carnei.	FLOWERS hanging down one over another all one way, forming a little head, of a pale red colour.
CALYX: PERIANTHIUM hexaphyllum, foliolis hirfutis, duo inferiora ovato-lanceolata, cætera linearia, *fig.* 2.	CALYX: a PERIANTHIUM of fix leaves, the leaves hairy, the two lowermoft of an oval-pointed fhape, the reft linear, *fig.* 2.
COROLLA ovata, monopetala, ore quadrifido, laciniis reflexis, *fig.* 3.	COROLLA oval, monopetalous, the mouth divided into four fegments, which turn back, *fig.* 3.
STAMINA: FILAMENTA octo, fubulata, alba, corollà breviora, receptaculo inferta; ANTHERÆ fagittatæ, conniventes, purpureæ, biforaminofæ, bicornes, *fig.* 4, 5, 6.	STAMINA: eight FILAMENTS, tapering, white, fhorter than the Corolla, inferted into the receptacle; ANTHERÆ arrow-fhaped, clofing together, purple, having two apertures for the difcharge of the Pollen, and two little horns, *fig.* 4, 5. 6.
PISTILLUM: GERMEN cylindraceum, fubfulcatum, villofum, glandulà ad bafin cinctum, *fig.* 7, 8; STYLUS filiformis, purpurafcens, *fig.* 9; STIGMA, obtufum, *fig.* 10.	PISTILLUM: GERMEN cylindrical, flightly grooved, villous, furrounded at bottom by a gland, *fig.* 7, 8; STYLE filiform, purplifh, *fig.* 9. STIGMA blunt, *fig.* 10.
PERICARPIUM: CAPSULA fubrotunda, villofa, apice truncata, quadrivalvis, *fig.* 11, 12.	SEED-VESSEL: a roundifh CAPSULE covered with a kind of down, cut off as it were at top, having four valves, *fig.* 11, 12.
SEMINA plurima, minuta, flavefcentia, *fig.* 13, 14.	SEEDS numerous, minute, and yellowifh, *fig.* 13, 14.

THIS fpecies of *Heath*, though not applicable to fuch a variety of ufes as fome of the others, is not inferior to any of them in the beauty and delicacy of its flowers, which in general are of a pale red colour, but fometimes they occur entirely white.

It is obvioufly enough diftinguifhed from the reft, not only by its flowers growing in a kind of pendulous clufter on the tops of the ftalks, but by its leaves alfo, which growing by fours on the ftalk, form a kind of crofs; thefe are edged with little ftiff hairs, each of which has a fmall globule at its extremity.

At the latter end of the Summer it contributes its fhare with the others to decorate and enliven thofe large tracts of barren land which too often meet the eye in many parts of this kingdom.

It delights to grow in a moifter fituation than fome of the others, and will thrive well enough in gardens, if taken up either in Spring or Autumn with a quantity of earth about its roots; this is neceffary, as the *Heaths* in general bear tranfplanting ill.

Erica tetralix.

POLYGONUM BISTORTA. THE GREATER BISTORT OR SNAKE-WEED.

POLYGONUM *Linnæi Gen. Pl.* OCTANDRIA TRIGYNIA.

    Raii Synopfis, Genus quintum. HERBÆ FLORE IMPERFECTO SEU STAMINEO, (VEL APETALO POTIUS.)

POLYGONUM *Biftorta* caule fimpliciffimo, monoftachyo, foliis ovatis in petiolum decurrentibus. *Linnæi Syft. Vegetab. p.* 311.

POLYGONUM radice lignofa contorta, fpica ovata, foliorum petiolis alatis. *Haller. Hift. v.* 2. 258.

COLUBRINA Seu Serpentaria fœmina. *Fufchii icon.* 774.

SERPENTARIA mas five Biftorta. *Fufchii icon.* 773.

BISTORTA major radice minus intorta. *Bauhin. Pin.* 192.

BISTORTA major radice magis intorta. *Bauhin. Pin.* 192.

BISTORTA major *Gerard emac* 379. *major vulgaris Parkinfon* 391. *Raii Synopfis* 147. *Hudfon. Fl. Angl.* 146. *Flor. Dan. Ic.* 421.

RADIX perennis, craffitie digiti, plus minufve intorta, externe caftanea, interne carnea, fibris et ftolonibus plurimis inftructa.

CAULIS pedalis aut bipedalis, *fimplex,* fuberectus, folidus, articulatus, (geniculi tumidi) teres, lævis.

STIPULÆ vaginantes, apice membranacæ, *marcefcentes, ore obliquo.*

FOLIA cordato-lanceolata, undulata, fubtus cærulefcentia, glabra, inferiora in petiolos decurrentia, fuperiora amplexicaulia in ftipulas definentia.

FLORES fpicati, fpica oblongo-ovata, denfa.

BRACTEÆ membranaceæ, marcefcentes, biflores, bivalves, valvula inferiore tricufpidata cufpide medio longiore quafi ariftata, flores pedicellati, pedicellis calyce longioribus.

CALYX five COROLLA fubovata, quinquepartita, carnea, laciniis ovatis, obtufis, concavis. *fig.* 1. 3.

STAMINA: FILAMENTA octo, fubulata, alba, corollâ longiora, ANTHERÆ biloculares, purpurafcentes, incumbentes. *fig.* 2.

PISTILLUM: GERMEN triquetrum, fanguineum, STYLI tres longitudine ftaminum; STIGMATA parva, rotunda. *fig.* 5. 6. 7.

NECTARIUM. glandulæ rubræ in fundo calycis, *fig.* 4.

SEMEN triquetrum, fufcum, mucronatum, nitens, vernice quafi obductum. *fig.* 8.

ROOT perennial, the thicknefs of one's finger, more or lefs crooked, externally of a chefnut, internally of a flefh colour, furnifhed with numerous fibres and creepers.

STALK from one to two feet high, *fimple,* nearly upright, folid, jointed, (the joints fwelled,) round and fmooth.

STIPULÆ enclofing the Stalk as in a fheath, at top membranous, *withered, the mouth oblique.*

LEAVES: *the bottom leaves* fomewhat heart fhaped and pointed; waved at the edges, fmooth, underneath blueifh and continued down the footftalks, the upper leaves embracing the ftalk, and terminating in the ftipulæ.

FLOWERS growing thickly in a fpike, the fpike of an oblong oval fhape.

FLORAL LEAVES membranous, and withered, containing two flowers and having two valves, the lower valve three pointed, the middle point running out into a kind of arifta or beard, the flowers growing on footftalks which are longer than the Calyx.

CALYX OR COROLLA, of an oval fhape and flefh coloured, divided into five fegments, which are oval, obtufe, and concave. *fig.* 1. 3.

STAMINA: eight FILAMENTS, tapering, white, and longer than the Calyx; the ANTHERÆ bilocular, purplifh, and laying acrofs the filaments. *fig.* 2.

PISTILLUM the GERMEN three fquare, of a deep red colour, three STYLES the length of the Stamina; the STIGMATA fmall and round. *fig.* 5. 6. 7.

NECTARIUM: feveral fmall red glands in the bottom of the Calyx. *fig.* 4.

SEED: triangular, brown, pointed, and fhining as it varnifhed. *fig.* 8.

WHEN a Plant not intended to be cultivated, in any refpect prevents the growth of one which is the object of Cultivation, fuch a plant, however beautiful, may with propriety be called a Weed; nor will the elegance or utility of the Biftort, fecure it in the eftimation of the Farmer, from that appellation.

This Plant generally grows in moift Meadows, and flowers in May and June; when it has once taken root, it propagates very faft, and frequently will form large patches, to the exclufion of a confiderable portion of the Grafs; nor is it deftroyed but with the greateft difficulty. Happily, our Farmers about Town are pretty much ftrangers to this Plant, as it is met with but rarely. It grows plentifully in a Meadow by the fide of *Bifhop's Wood* near *Hampftead,* and my obliging Friend Dr. ALLEN informs me he has found it about *Batterfea.*

As an aftringent Medicine, the Biftort appears to poffefs confiderable virtue, and as fuch may with propriety be made ufe of in all cafes where aftringents are required; but more particularly in long continued evacuations from the Bowels, and other difcharges both ferous and fanguineous. It is recommended alfo to faften teeth which are loofe, and may be ufed either in powder, infufion, or extract. If it could be procured in fufficient quantity to make it anfwer, it might well be applied to the purpofe of tanning Leather.

In fome parts of England the leaves are eat as a Pot-herb.

_Polygonum_Bistorta_

Polygonum Persicaria

POLYGONUM PERSICARIA. COMMON SPOTTED PERSICARIA.

POLYGONUM *Linnæi Gen. Pl.* OCTANDRIA TRIGYNIA.

Raii Syn. Gen. 5 HERBÆ FLORE IMPERFECTO SEU STAMINEO, VEL APÉTALO POTIUS.

POLYGONUM *Perficaria* floribus hexandris femidigynis, pedunculis lævibus, ftipulis ciliatis, fpicis ovato-oblongis erectis.

POLYGONUM *Perficaria* floribus hexandris digynis, fpicis ovato-oblongis, foliis lanceolatis, ftipulis ciliatis, *Lin. Syft. Vegetab. p.* 312. *Flor. Suecie. p.* 130.

POLYGONUM foliis ovato-lauiceolatis, fubhirfutis, fpicis ovatis, vaginis ciliatis. *Haller. hift. Helv. v. 2. p.* 257.

PERSICARIA mitis maculofa et non maculofa. *Bauhin. Pin. p.* 101.

PERSICARIA maculofa *Gerard. emac.* 445. vulgaris mitis feu maculofa. *Parkinfon.* 856. *Raii Syn. ed. 3. p.* 145. *n.* 4. Dead or fpotted Arfmart. *Hudfon Flor. Angl. p.* 147. *n.* 4. *Scopoli Fl. Carniol p.* 279.

RADIX fimplex, fibrofa.	ROOT fimple and fibrous.
CAULIS erectus, ad bafin aliquando repens, pedalis ad tripedalem, ramofus, teres, glaber, ad geniculos fenfim incraffatus, fæpe rubens : fub geniculis puncta radicalia difcernantur quamvis huic fpeciei non propria.	STALK upright, fometimes creeping at bottom, from one to three feet high, branched, round, fmooth, gradually thicker at the joints, often of a red colour : a little beneath each joint fome radical points are obfervable, which however are not peculiar to this fpecies.
RAMI alterni, e fingulo geniculo prodeuntes, patentes, fæpe diffufi.	BRANCHES alternate, proceeding from each joint, fprending, frequently very much fo.
STIPULÆ vaginantes, liquore vifcido fæpe repletæ, ciliatæ.	STIPULÆ embracing the ftalk, frequently full of a vifcid liquid, and terminated by long ciliæ or hairs.
FOLIA lanceolata, fubpetiolata, margine nervoque medio fubhirfutis, utrinque lævia, maculâ ferrum equinum quodammodo referente fæpius notata.	LEAVES lanceolate, with fhort foot-ftalks, the edge and midrib flightly hairy, fmooth on both fides, in general having a large fpot on the middle of the leaf fomewhat like a horfe fhoe.
PEDUNCULI læves.	FOOT-STALKS of the flowers, fmooth.
FLORES fpicati, rofei, Spicæ terminales, erectæ, fubovatæ.	FLOWERS growing in fpikes, of a bright rofe colour, the fpikes terminal, upright, of a fomewhat oval fhape.
CALYX: PERIANTHIUM quinquepartitum, coloratum, perfiftens, fegmentis ovatis obtufis, *fig.* 1, 2.	CALYX : a PERIANTHIUM divided into five fegments, coloured, and perfifting, the fegments oval and obtufe, *fig.* 1, 2.
COROLLA nulla.	COROLLA wanting.
STAMINA : FILAMENTA fex fundo calycis inferta longitudine corollæ ANTHERÆ rubentes, *fig.* 2.	STAMINA: fix FILAMENTS inferted into the bottom of the Calyx, the length of the Corolla; the ANTHERÆ redifh, *fig.* 2.
PISTILLUM: GERMEN ovatum, compreffum, aut triquetrum, *fig.* 3, 6. STYLUS *ad medium ufque bifidus* fæpe trifidus, *fig.* 5, 6. STIGMATA duo aut tria fubrotunda, *fig.* 4, 7.	PISTILLUM: GERMEN oval and flat, or three-fquare, *fig.* 3, 6. STYLE *divided down to the middle into two,* often into three parts, *fig.* 5, 6. STIGMATA two or three, and round, *fig.* 4, 7.
SEMEN unicum, nitidum, aut fubovatum, acuminatum, ad unum latus leviter convexum ; *fig.* 9, 11, aut trigonum, *fig.* 10, 12.	SEED one, fhining, either of an oval pointed fhape and flightly convex on one fide, *fig.* 9, 11. or three-fquare, *fig.* 10, 12.

The very great fimilarity which exifts between the feveral fpecies of the Polygonums, has occafioned no fmall degree of trouble to Botanifts, in rightly afcertaining the limits of each Species and Variety ; a difficulty not to be overcome while Books are confulted more than Nature. Senfible of the truth of this obfervation, and earneftly defirous of arriving at fome certainty on this fubject, we have examined a vaft number of all the different Species and Varieties of Polygonum which our neighbourhood affords, compared them with one another, fown the feeds, and cultivated many of them; and if we do not deceive ourfelves, have reduced fome of the more difficult ones to their true Species and Varieties.

As what we relate concerning thefe plants is no more than the refult of the moft accurate and repeated inveftigation, affifted by the microfcope, we fhall be the lefs concerned becaufe we differ from Authors of the moft refpectable Authority.

The writer who gives an account of all the known plants in the univerfe, cannot be fuppofed to have the opportunity of being fo minute in his enquiries as one who defcribes the plants of a particular fpot, which as they grow are conftantly the objects of his attention.

We have ventured to alter Linnæus's Specific defcription of this plant, which ftands thus.

Polygonum floribus hexandris digynis, fpicis ovato-oblongis, foliis lanceolatis, ftipulis ciliatis. to

Polygonum floribus hexandris femidigynis, pedunculis lævibus, ftipulis ciliatis, fpicis ovato-oblongis erectis.

We have not made this alteration from an idle defire of differing from fo great a Man, whom we truly refpect and revere, but folely to make the diftinctions betwixt thofe plants more obvious, and thereby add our mite to the general ftock of Botanic knowledge. In fpecific defcriptions, the diftinguifhing marks fhould as much as poffible be contrafted or oppofed to each other, in thefe plants this does not feem to have been fufficiently attended to. What we have principally in view by altering the Specific defcription is to diftinguifh it from the *Polygonum Penfylvanicum* and its varieties, of which there are feveral, and to which the *Polygonum Perficaria* in its general habit is exceeding nearly allied.

In all the flowers of this Species which we have examined, the Style has been divided juft *half way down,* hence we have called the flowers *Semidigyni,* had it been divided down to the bafe they would with propriety have been called *Digyni.* In moft of the flowers the Style is divided into two parts, and the Germen is a little convex on each fide, in fome of the flowers the ftyle is divided into three, hence thofe flowers might be called *Semitrigyni,* and when this is the cafe the Germen is always triangular. In the *Polygonum Penfylvanicum* the Style is divided *nearly to the bafe,* this difference then in the divifion of the Style, is of confiderable confequence in diftinguifhing the two Species and their varieties from each other.

The footftalks which fupport the flowers in this Species, are quite fmooth, in the *Polygonum Penfylvanicum,* they are befet with a great number of minute glands, which gives them a manifeft roughnefs, and contributes to characterife that Species.

The Stipulæ are furnifhed with long Ciliæ or Hairs, particularly towards the top of the plant, in the *Polygonum Penfylvanicum* thefe are wanting. Thefe two plants likewife differ much in the form of their feeds, of which we fhall fpeak more fully in our account of the latter.

The flowers always grow in upright fpikes of an oval fhape more or lefs round; by thefe two characters this Species is at once diftinguifhed from the *Polygonum Hydropiper,* the fpikes of which are *filiform and pendulous.*

The leaves are moft commonly fpotted, but this is neither conftant nor peculiar to this Species, and difference of fize only forms the principle variety to which it is fubject.

It grows exceedingly common in all our Ditches, and flowers in Auguft and September; its bloffoms are beautiful and laft a confiderable time, was it not fo common, it would probably be thought worthy of a place in our Gardens.

No particular virtues or ufes are attributed to it.

POLYGONUM   PENSYLVANICUM.   PALE - FLOWERED

PERSICARIA.

POLYGONUM *Linnæi Gen. Pl.* OCTANDRIA TRIGYNIA.

Raii Syn. Gen. 5. HERBÆ FLORE IMPERFECTO SEU STAMINEO (VEL APETALA POTIUS.)

POLYGONUM floribus hexandris, digynis; ſtipulis muticis; pedunculis ſcabris; ſeminibus utrinque depreſſis.

POLYGONUM floribus octandris digynis, pedunculis hiſpidis, foliis lanceolatis, ſtipulis muticis.

Linnæi Syſt. Vegetab. Sp. Plant. p. 519.

PERSICARIA mitis major foliis pallidioribus. *D. Bobarti, Dead Arſmart the greater with pale leaves.*

Raii Syn. ed. 3. *p.* 145. *Hudſon Fl. Angl. p.* 148.

RADIX fibroſa, annua.

ROOT fibrous and annual.

CAULIS tripedalis circiter, teres, glaber, fiſtuloſus, ramoſus ; rami patentes, geniculis maxime incraſſatis.

STALK about three feet high, round, ſmooth, hollow, branched, and the branches ſpreading, and the joints very much ſwelled.

FOLIA ovato-lanceolata, ſupra glabra, ſubtus glandulis punctata, ſæpe pubeſcentia, ciliata, nunc maculata nunc immaculata.

LEAVES of an oval pointed ſhape, ſmooth on their upper ſurface, underneath dotted with ſmall glands, and often downy, edged with little hairs, ſometimes with and ſometimes without ſpots.

PETIOLI ſubtus hirſuti, ſcabriuſculi.

FOOT-STALKS of the leaves hairy underneath, with a ſlight roughneſs to the touch.

STIPULÆ baſi nervoſæ, muticæ.

STIPULÆ rib'd at bottom, and not terminated by any hairs.

PEDUNCULI pilis brevibus glanduliferis ſcabri. *fig.* 1.

FOOT STALKS of the flowers rough with little glands. *fig.* 1.

FLORES herbacei, pedunculis brevibus inſidentes, denſe glomerati, ſpicæ ovatæ, ſeminibus maturis ſubnutantes.

FLOWERS of a greeniſh colour, ſitting on ſhort foot-ſtalks, and growing thickly together ; ſpikes oval, and when the ſeeds are ripe drooping a little.

CALYX: PERIANTHIUM quinquepartitum, laciniis ovatis, obtuſis, *fig.* 2, 3.

CALYX : a PERIANTHIUM divided into five ſegments, which are oval and obtuſe, *fig.* 2, 3.

COROLLA nulla.

COROLLA wanting.

STAMINA : FILAMENTA ſex, ſubulata, alba, Corollâ paulo breviora ; ANTHERÆ biloculares ; POLLEN globoſum, *fig.* 4.

STAMINA : ſix FILAMENTS, tapering, white, a little ſhorter than the Corolla ; ANTHERÆ bilocular ; POLLEN globular, *fig.* 4.

PISTILLUM : GERMEN ſubovatum ; STYLUS fere ad baſin diviſus ; STIGMATA duo ſubrotunda, *fig.* 5, 6.

PISTILLUM : GERMEN ſomewhat oval; STYLE divided nearly down to the baſe ; STIGMATA two, roundiſh, *fig.* 5, 6.

SEMEN cordatum, acuminatum, compreſſum, *medio depreſſum*, nitidum, *fig.* 9, 10, magnit. nat. *fig.* 7, 8, lente auct. ſubinde obtuſe triquetrum, *fig.* 12.

SEED heart-ſhaped, pointed, flat, *with a depreſſion in the middle*, ſhining, *fig.* 9, 10, of its natural ſize, *fig.* 7, 8, magnified, ſometimes obtuſely triangular, *fig.* 12.

The plant here figured, is the *Perſicaria mitis major foliis pallidioribus, D. Bobarti,* and which is particularly deſcribed in the 3d. edition of RAY's *Synopſis, p.* 145 : from the conſquancy of this deſcription, with that which LINNÆUS had given of the *Polygonum Penſylvanicum,* in the 3d. edition of his *Species Plantarum,* Mr. HUDSON ſet it down in his *Flora* as that ſpecies : and LINNÆUS, in the laſt edition of his *Syſtema Vegetab.* as a confirmation of our Engliſh *Polygonum's,* being the ſame with his *Penſylvanicum,* quotes BOBARTS's deſcriptive name.

By RAY, LINNÆUS, and HUDSON, then, it is made a diſtinct ſpecies ; by HALLER it is conſidered as a variety of the *Polygonum Perſicaria* ; but as the Baron forms his judgment from dried ſpecimens that were ſent him, in which many of the diſtinguiſhing characters of this plant would be unavoidably loſt, he ſeems the moſt likely to be miſtaken : I ſhall therefore join in making it a diſtinct ſpecies ; and I truſt ſhall give ſuch ſtriking additional characters, as will ſettle this matter beyond diſpute.

The true *Polygonum Penſylvanicum* (for there are ſeveral varieties of it) has the greateſt affinity with the *Polygonum Perſicaria,* but differs from it in the following particulars, viz. place of growth, ſize, ſtipulæ, leaves, foot-ſtalks of the leaves, foot-ſtalks of the flowers, ſtyle, and ſeeds.

While the *Polygonum Perſicaria* uſually delights to grow by the ſides of moiſt ditches, the *Penſylvanicum* prefers a richer and more luxuriant ſoil ; and ſo common is it with us about town, that there is ſcarce a dunghill on which it may not be found : indeed in its attachment to this particular ſoil, it reſembles many of the *Chenopodiums* or *Oraches.* Was it never to occur in other ſituations, ſome might be ready to ſuſpect that it was a variety of the *Perſicaria* ariſing from richneſs of ſoil ; but it is frequently found in other places : and I remember once to have ſeen the *Polygonum Perſicaria, Hydropiper,* and *Penſylvanicum,* all growing by the ſide of a ſtream within ſix inches of each other.

In its moſt common ſtate it is much larger than the *Polygonum Perſicaria,* and its joints in particular are more ſwelled ; its Stipulæ are much more ſtrongly rib'd at bottom, and have no Ciliæ ; its leaves are broader, the veins ſomewhat deeper, and more ſtrongly marked ; the hairs on the edges of the leaves more viſible, but particularly ſo under the foot-ſtalk of the leaf, to which they give a manifeſt roughneſs : in the uppermoſt leaves the under ſide is generally dotted with very minute glands, while in the lowermoſt it is covered with a kind of down : this laſt character, though contrary to what LINNÆUS aſſerts, is never ſeen in the *Polygonum Perſicaria* ; but in this ſpecies it is always more or leſs predominant. The foot-ſtalks of the flowers are thickly beſet with little yellowiſh glands, ſtanding on ſhort foot-ſtalks, which ſometimes extend half way down the plant ; this appearance never or exceeding rarely occurs in the *Polygonum Perſicaria :* the flowers are of a pale or greeniſh hue, and form thicker and larger ſpikes than in the *Polygonum Perſicaria,* and when ripe are ſo heavy as frequently to hang down a little : the Style is divided very nearly down to the Germen, while in the *Polygonum Perſicaria* it is divided only half way ; and *this diviſion of the Style,* I look upon as one of the moſt conſtant and certain criteria of this ſpecies : laſtly the form of the ſeeds contributes not a little to the farther aſcertaining and fixing it ; in the *Perſicaria* the ſeeds are either triangular, or of a pointed oval ſhape, with a little *convexity* on each ſide ; in this ſpecies it is in general flat, with a *depreſſion* on each ſide ; it is alſo larger and broader ; now and then a ſeed occurs forming an unequal triangle, but theſe are very rare, while the triangular ſeed is moſt frequent in the *Polygonum Perſicaria.*

Polygonum Pensylvanicum

Polygonum Pensylvanicum var: caule maculato.

Polygonum Pensylvanicum. var. caule maculato.

Spotted-stalk'd Persicaria.

PERSICARIA latifolia geniculata, caulibus maculatis. *D. Rand. Raii Syn. p.* 145.

PERSICARIA maculosa procumbens foliis subtus incanis. *Raii Syn. p.* 146. eadem est planta solo
autem minus læto proveniens.

Such then is the difference, which from repeated examinations, I have been able to discover betwixt the *Polygonum Persicaria* and the *Pensylvanicum* in its most common state ; in this state however it does not always occur, but is subject to more Varieties than any of our other *Persicaria's* : without any desire of multiplying them, I make the following, having found them all about *London* :

1 *Polygonum Pensylvanicum. var.*   *caule et floribus rubris.*
2 . . . . *caule maculato.*
3 . . . . *foliis subtus incanis.*

The first of these varieties is very often found with the true species on dunghills, as also in corn-fields, and is like it in every respect excepting its colour, the stalks and flowers being red, but not so beautifully bright as those of the *Polygonum Persicaria.*

The second variety here figured, which indeed comes near to a distinct species, grows much in the same situations, and oftentimes with the *Polygonum Persicaria* in the ditches about *St. Georges-fields*, particularly in a large ditch on the right-hand side of the road between the end of *Blackman-Street* and *Newington*, where it is very common in the month of *September*. It not only differs from the other in having its stalk spotted with red, a character which it keeps very constantly, but its spikes are much slenderer, rather more so even than those of the *Persicaria*, of a red colour, but not quite so bright as those of that plant : the under side of the foot-stalk of the leaves is remarkably rough ; the little glands on the foot-stalks of the flowers, and the parts of the fructification are similar to those of the true species, but the seeds are smaller : when this variety grows in the rich soil abovementioned, it is full as large as the *Pensylvanicum* itself ; but when it grows in a different soil and situation, as on the watery parts of *Blackheath* and *Peckham-Rye*, it becomes much smaller, generally has its leaves white underneath, and will certainly be taken for the *Polygonum Persicaria* if not attentively examined : its spotted stalk and the roughness of the foot-stalks of the leaves will however readily discover it.

The third variety, with leaves hoary on the under side, is found here and there in corn-fields and other places, where the soil is not very rich, and is obviously enough distinguished.

Besides these striking varieties, it is subject, like all other plants, to vary in size according to the richness or poverty of the ground on which it grows, and like the *Polygonum Persicaria*, its leaves are sometimes spotted and sometimes not.

This descriptive account will perhaps appear tedious and uninteresting to some ; if however by these practical observations, the obscurity which has hitherto dwelt on this difficult Genus, shall in some degree be removed, and the road of investigation made easier to the young Botanist, I shall think my time usefully employed ; I would not however wish him to take upon trust what is here advanced, but to examine each plant and its several parts for himself ; thus he will become improved, and be able perhaps to throw a still greater light on the subject.

The Sparrow and other small Birds are very fond of the seeds of this species and its varieties ; but the Farmer should carefully weed them from his dunghills.

Polygonum Hydropiper. Biting Persicaria or Water Pepper.

POLYGONUM *Linnæi Gen. Pl.* Octandria Trigynia.

Cal. o. *Cor.* 5-partita, calycina. *Sem.* 1, angulatum.

Rai Syn. Gen. Herbæ flore imperfecto seu stamineo vel apetalo potius.

POLYGONUM *Hydropiper* floribus hexandris semidigynis; foliis lanceolatis, undulatis, immaculatis; spicis filiformibus nutantibus.

POLYGONUM *Hydropiper* floribus hexandris semidigynis, foliis lanceolatis, stipulis submuticis. *Linn. Syst. Vegetab. p.* 312.

POLYGONUM foliis ovato lanceolatis, spicis florigeris, vaginis calvis. *Haller. hist. p.* 256. *n.* 1554.

POLYGONUM *Hydropiper*. *Scopoli Fl. Carniol. n.* 467.

PERSICARIA urens seu Hydropiper. *Bauhin. pin.* 101.

PERSICARIA vulgaris acris seu minor. *Parkinson.* 856.

HYDROPIPER. *Gerard. emac.* 445. *Rai Syn. p.* 144. Water-pepper, Lakeweed or Arsmart. *Hudson Fl. Angl. p.* 148.

RADIX annua, fibrosa.

CAULIS erectus, ramosus, basi nonnunquam repens, pedalis ad tripedalem, geniculis incrassatis, demum ruberrimus.

FOLIA lanceolata, *undulata*, e viridi flavescentia, glabra.

STIPULÆ ciliatæ.

FLORES spicati, *spicæ tenues, demum nutantes.*

CALYX: Perianthium quadripartitum, *glandulis minimis adspersum*, laciniis obtusis, concavis, *fig.* 1, 2, 3.

COROLLA nulla.

STAMINA: Filamenta sex alba; Antheræ albæ biloculares, *fig.* 3.

PISTILLUM: Germen ovatum; Stylus bifidus, ad medium usque divisus; Stigmata duo, rotunda, *fig.* 4, 5.

SEMEN ovato-acuminatum, castaneum, *fig.* 6.

ROOT annual and fibrous.

STALK upright, branched, sometimes creeping at bottom, from one to three feet high, the joints swelled, finally becoming very red.

LEAVES lanceolate, *waved*, of a yellowish green colour and smooth.

STIPULÆ ciliated.

FLOWERS growing in spikes, *which are slender and finally drooping.*

CALYX: a Perianthium divided into four segments, *sprinkled with very minute glands*, the segments blunt and hollow, *fig.* 1, 2, 3.

COROLLA wanting.

STAMINA six white Filaments; Antheræ white and bilocular, *fig.* 3.

PISTILLUM: Germen oval; Style bifid, divided down to the middle; two round Stigmata, *fig.* 4, 5.

SEEDS of an oval pointed shape, and chesnut colour, *fig.* 6.

It is one of the maxims laid down by the Author of that system of Botany which at present is so deservedly held in esteem, and which I trust for the sake of this delightful science will for ever withstand the attempts of all those who frame systems merely to raise themselves into consequence, that in all specific descriptions taste is to be excluded: some may perhaps be ready to treat this as too dogmatical, but when they come to find that both the *Hydropiper* and *Sedum acre*, plants which in general are very hot and biting, sometimes are found insipid, they will readily adopt it as founded in strict propriety.

The present species of Polygonum very properly receives its name of *Hyaropiper* from its hot and biting taste, which appears to arise from its essential oil dispersed in little cells or glands all over the plant, but more particularly observable on the Calyx with a small magnifier, and which, if tasted, will be found to be more biting than any other part of the plant: this quality which is peculiar to the *Hydropiper*, generally leaves a strong Idea of the plant on the mind of the Tyro: but it is has other more invariable characters whereby it may be distinguished. Notwithstanding its obvious difference from the other plants of this genus, apparent even to such as know very little of Botany, both Scopoli and Haller seem to entertain doubts whether it be really distinct from the P. *Persicaria* and P. *minus*.

The three plants as they usually grow, and I have seen them all three grow together, are certainly distinct enough: but there are some intermediate varieties which bring them very near together, and perhaps justify such suspicions: a variety of the *Hydropiper*, scarce differing in any other respect but its insipidity, I have now and then met with in the same situation as we usually find the true species: from the P. *Persicaria* it differs principally in its leaves, spikes, form and size of its seeds; and first its leaves are of a yellower hue, more undulated, and never marked with any spots; its spikes are slender, and when the seeds are ripe they bend and hang down; the seeds are much larger, more acuminated, and of a chesnut colour; its stipulæ are very evidently ciliated; though Haller makes their want of ciliæ one of its striking characters; and Linnæus also calls them *submuticæ*, which certainly tends to mislead.

It is the only *Persicaria* that has any pretensions to be an active medicine: given in infusion or decoction it proves diuretic, hence it is made use of in the Dropsy and Jaundice; and the distilled water of it is recommended by Boyle as efficacious in the Stone and Gravel: Linnæus informs us that the plant will dye Woolen cloth of a yellow colour.

Although the herb is so acrid, the seeds are insipid and nutritive.

It is found in great abundance in all those places which lie under water during the Winter, flowers in September, generally a month later than the P. *Persicaria*: in exposed places it becomes very red in going off.

Polygonum Hydropiper.

Polygonum aviculare.

Polygonum Aviculare. Birds Polygonum or Knot-Grass.

POLYGONUM *Linnæi Gen Pl.* OCTANDRIA TRIGYNIA.

Cal. o. *Cor.* 5-*partita, Calycina. Sem.* 1. *angulatum.*

Raii Syn. Gen. 5. HERBÆ FLORE IMPERFECTO SEU STAMINEO. (VEL APETALA POTIUS.)

POLYGONUM *aviculare* floribus octaudris trigynis axillaribus, foliis lanceolatis, caule procumbente herbaceo. *Linn. Syft. Vegetab. p.* 312. *Sp. Pl.* 519. *Fl. Suecic. n.* 339.

POLYGONUM procumbens, foliis linearibus, acutis, floribus folitariis. *Haller hift. n.* 1560.

POLYGONUM *aviculare. Scopoli Fl. Carniol. n.* 471.

POLYGONUM mas vulgare. *Gerard emac.* 451.

POLYGONUM mas vulgare majus. *Parkinfon* 443.

POLYGONUM fou Centinodia. *I. Baubin* 3. 374. *Raii Syn. p.* 146. *Hudfon Fl. Angl. p.* 149.

RADIX annua, fimplex, lignofa, multis fibris donata, terram firmiter apprehendeus ut extirpatu difficilis fit, fapore adftringente.

ROOT annual, fimple, woody, furnifhed with many fibres, taking ftrong hold of the earth, fo as to be with difficulty pulled up, and of an aftringent tafte.

CAULES plures, plerumque procumbentes, interdum vero fuberecti, dodrantales, ramofi, tenues, ftriati, læves, teretes, geniculati, ad geniculos paululum incraffati.

STALKS feveral, generally procumbent, fometimes nearly upright, about nine inches in length, branched, flender, ftriated, fmooth, round, jointed, the joints a little fwelled.

FOLIA quam maxime variantia, ovata, lanceolata aut etiam linearia, alterna, lævia, e vaginis ftipularum prodeuntia.

LEAVES varying exceedingly, oval, lanceolate, or fometimes even linear, alternate, fmooth, proceeding from the fheaths of the Stipulæ.

STIPULÆ vaginantes, membranaceæ, albidæ, nitidæ, apice fibrofæ.

STIPULÆ forming a fheath round the joints, membranous, white, fhining, at top fibrous.

FLORES axillares, e vaginis ftipularum cum foliis prodeuntia.

FLOWERS axillary, proceeding with the leaves from the fheaths of the Stipulæ.

CALYX : PERIANTHIUM quinquepartitum, laciniis ovatis concavis, patentibus, dimidio inferiore viridi, fuperiore albo, faepe colorato, *fig.* 1, 2.

CALYX : a PERIANTHIUM divided into five fegments, the laciniæ oval, concave and fpreading, the lower half green, the upper half white and often coloured, *fig.* 1, 2.

COROLLA nulla.

COROLLA wanting.

STAMINA : FILAMENTA octo corolla breviora ; ANTHERÆ flavæ, *fig.* 2, auct.

STAMINA : eight FILAMENTS fhorter than the Corolla, ANTHERÆ yellow, *fig.* 2, magnified.

PISTILLUM : GERMEN triquetrum ; STYLUS longitudine ftaminum, trifidus ; STIGMATA tria, rotunda, *fig.* 3, auct.

PISTILLUM : GERMEN triangular ; STYLE the length of the Stamina, trifid ; STIGMATA three, round, *fig.* 3, magnified.

SEMEN triquetrum, nigricans, intra calycem, *fig.* 4.

SEED triangular, of a blackifh colour, contained within the Calyx, *fig.* 4.

Thofe plants which have been obferved to be eaten by cattle, have often obtained the name of *Grafs*, although they have not poffeffed the leaft fimilitude to thofe which are real Graffes, and the prefent plant is one of thefe. Cattle in general are fond of it, and hogs in particular eat it with great avidity. The feeds afford fuftenance to many of the fmall birds, whence it has acquired the name of *aviculare*. The Caterpillar of the *Phalæna rumicis* (with us the *Knot-grafs Moth*,) I have frequently found feeding on its leaves, although it is by no means confined to this plant : in Sweden, LINNÆUS informs us it feeds on the *Dock* (*Rumex*,) and *Sow-thiftle*.

This fpecies of *Polygonum* may be confidered as one of our moft common plants ; it delights to grow in a fandy or gravelly foil, on banks, and by the fides of roads and paths, being of quick growth, and fpreading a great deal of ground ; it often covers whole fields, that by turning in of cattle, have had their natural coat of grafs deftroyed.

Where a plant of this fpecies happens to grow fingly in a rich foil, it will often cover the fpace of a yard or more in diameter, and the leaves become broad, and large ; but when it grows very thick together, by the fides of paths, it is in every refpect fmaller, and the ftalks are more upright. It is fubject, like moft other plants, to feveral varieties, and of thefe are the *Polygonum brevi anguftoque folio*, and the *Polygonum oblongo anguftoque folio* of C. Baubine.

It has been confidered by antient writers, as poffeffing fome medical virtue, particularly as an Aftringent, and is by them recommended in Diarrhæas, Dyfenteries, Bleeding at the nofe, and other Hemorrhages ; but in the prefent practice, its ufe feems juftly fuperfeded by more efficacious medicines.

* 174. Linn. Faun. Suecic. p. 318. n. 1200. Rœfel. cl. 1. Pap. N.H. 1. 27. Albin Infect. pl. 11.

Polygonum minus.

Polygonum minus. Small, Creeping, Narrow-leaved

Persicaria.

POLYGONUM *Linnæi Gen. Pl.* Octandria Trigynia.

Cal. o. Cor. 5-partita calycina. Sem. 1, angulatum.

Raii Syn. Gen. 5. Herbæ flore imperfecto seu stamineo vel apetala potius.

POLYGONUM *minus* floribus hexandris, fubmonogynis, foliis lineari-lanceolatis, caule bafi repente.

POLYGONUM *minus* hexandris digynis foliis lanceolatis, ftipulis ciliatis, caule divaricato patulo. *Hudfon Fl. Angl. p.* 148.

POLYGONUM foliis ovato-lanceolatis, glabris, fpicis ftrigofis, vaginis ciliatis. *Haller. hift. p.* 257. n. 1555.

PERSICARIA minor. *Bauhin Pin.* 1014? anguftifolia. *Bauhin Pin.* 101. 3?

PERSICARIA pufilla repens *Ger. emac.* 446. *Parkinfon.* 857. *Raii Syn.* 145. 2. Small Creeping Arfmart.

PERSICARIA anguftifolia ex fingulis geniculis florens. *Mer. Pin.* 90? *Raii Syn.* 145. 3. Narrow-leaved Lakeweed.

RADIX annua, fibrofa.

CAULES plures, dodrantales, aut pedales, *bafi repentes,* demum fuberecti, geniculati, (geniculis paululum incraffatis,) læves, rubicundi.

FOLIA *lineari-lanceolata,* pene avenia, fuperne glabra.

STIPULÆ vaginantes, ciliatæ.

SPICÆ tenues, parum nutantes, e fingulis geniculis prodeuntes.

CALYX : Perianthium quinquepartitum, perfiftens, coloratum, laciniis obtufis concavis, *fig.* 1.

COROLLA nulla.

STAMINA : Filamenta fex ; Antheræ biloculares, albæ, intra Corollam.

PISTILLUM : Germen ovatum aut triangulare ; Stylus filiformis, apice bifidus aut trifidus ; Stigmata duo aut tria rotunda, reflexa, *fig.* 2, 3.

SEMEN aut ovato-acutum aut triangulare, caftaneum, magnitudinis fere et formæ feminis Polygoni Perficariæ, *fig.* 4, 5.

N. B. Omnes partes fructificationis lente augentur.

ROOT annual, and fibrous.

STALKS feveral, about nine inches or a foot high, *creeping at bottom,* then becoming nearly upright, jointed, (the joints fomewhat thickened,) fmooth, of a reddifh colour.

LEAVES *betwixt linear and lanceolate,* fcarcely any appearance of veins, on thier upper furface fmooth.

STIPULÆ forming fheaths round the joints, and ciliated.

SPIKES flender and a little drooping, proceeding from each joint of the ftalk.

CALYX : a Perianthium divided into five fegments, which are obtufe and hollow, *fig.* 1.

COROLLA wanting.

STAMINA fix Filaments ; Antheræ bilocular, and white, within the Corolla.

PISTILLUM : Germen oval or triangular ; Style filiform, at top blifd or trifid ; Stigmata two or three, round and turned back, *fig.* 2, 3.

SEEDS oval or triangular, of a chefnut colour, nearly of the fame fize and fhape as the Polygonum Perficaria, *fig.* 4, 5.

N. B. All the parts of the fructification are magnified.

If the opportunity of feeing this plant growing wild had ever occured to the celebrated Swedifh Botanift, he would doubtlefs have confidered it as a diftinct fpecies ; at prefent he has placed it in the laft edition of his works, the *Syftema Vegetabilium,* as a variety of the *Polygonum Perficaria,* probably mifled by dried fpecimens of the plant : thofe who truft to fuch are exceeding liable to deceive both themfelves and others, particularly in plants whofe parts of fructification (from which it is fometimes neceffary to draw fpecific differences) are very minute—thofe in the living plants are with difficulty enough diftinguifhed, and in dried fpecimens not to be investigated.

Whoever has obferved the appearance which the *Polygonum minus* and *Perficaria* ufually put on, muft have been ftruck with the great diffimilarity of the two in their general habits ; and if they have taken the pains to examine the parts of fructification, they will, I am perfwaded, be convinced that both Mr. Ray and Hudson are juftifiable in making them diftinct fpecies.

It differs from the *Polygonum Perficaria* in its fize, growth of its ftalk, fhape of its leaves, form of its fpikes, and divifion of its Piftillum. In height it feldom exceeds a foot, whereas the *Perficaria* often occurs a yard high ; the ftalk of this fpecies creeps at bottom, in the *Perficaria* it never does : it is true in the *Perficaria,* and moft of the *Polygonums,* a number of little roots pufh themfelves out at the joints, which are next the ground ; but in this fpecies the ftalk at bottom is abfolutely procumbent, whilft in the *Perficaria* it is always upright ; the leaves are much narrower, approaching rather to linear than lanceolate, and on their upper furface have much lefs appearance of veins, than in the *Perficaria* ; the fpikes, inftead of being oval or nearly round, and upright, as in the *Perficaria,* are flender and a little drooping : the Piftillum, which is a part of very great confequence in determining many of the fpecies and varieties of this genus, is flightly divided at top only ; while that of the *Perficaria* is divided half way down ; hence as I have called that fpecies *femitrigynous,* I have called this *fubmonogynous.*

Hitherto I have met with this plant growing wild no where but in *Tothill-fields,* Weftminfter, where it makes ample amends for its fcarcity elfewhere, being found in the greateft abundance in the watry parts of thofe fields, along with the *Sifembrium fylveftre* in the month of September, when it is in full bloom.

At prefent it does not appear that it has any thing more than its fcarcity to recommend it to our notice.

Butomus umbellatus. Flowering Rush,
or WATER GLADIOLE.

BUTOMUS *Linnæi Gen. Pl.* Enneandria Hexagynia.

Raii Gen. 17. Herbæ Multisiliquæ feu Corniculatæ.

BUTOMUS umbellatus. *Lin. Spec. plant.* 532.

JUNCUS floridus major. *Bauhin. pin.* p. 12.

GLADIOLUS paluftris Cordi *Gerard. emac.* 29.

Raii Synopfis ed. 3. 273. *Hudfon. Fl. Angl.* 152. *Scopoli Fl. Carn. ed.* 2. p. 283. *Halleri hift. pl. Helv. Vol.* 2. 81.

RADIX perennis, alba, tuberculofa, tranfverfa, edulis ? ex inferiore parte radiculas præ)longas dimittens.

ROOT perennial, white, knobby, tranfverfe, eatable ? from its under fide fending down a great number of very long fibres.

SCAPUS pedalis ad orgyalem, teres, glaber.

STALK round, fmooth, from one to five or fix feet high, according to its place of growth.

FOLIA triquetra, fpongiofa, *fig.* 1. fcapo breviora ; ad bafin fpathacea, apicibus compreffis, tortuofis.

LEAVES triangular, fpongy, *fig.* 1. fhorter than the ftalk, at bottom fheathy, at top flat, and twifted.

FLORES in Umbella, ad triginta ; pedunculi digitales, e vaginis membranaceis prodeuntes.

FLOWERS numerous, to thirty, each on a fingle peduncle of about a finger's length, forming an Umbell, furrounded at bottom by withered membranous fheaths.

CALYX. Involucrum triphyllum, foliolis lanceolatis, marcefcentibus.

CALYX. An Involucrum of three leaves, fpear fhaped and withered.

COROLLA. Petala fex inæqualia, fubrotunda, concava, rofea, *fig.* 2, alternis minoribus, acutioribus. *fig.* 3.

COROLLA. Compofed of fix Petals, which are roundifh, concave, and moft commonly of a bright red ; *fig.* 2. the three exterior fmaller, and more pointed, *fig.* 3.

STAMINA. Filamenta novem, fubulata, *fig.* 4. 5. Antheræ infidentes, dum pollinem involvunt oblongæ, rubræ, quadrifulcatæ, mucrone brevi terminatæ, *fig.* 6. 7, emiffo polline fubcordatæ, compreffæ, bilamellofæ, *fig.* 4. Pollen flaviffimum.

STAMINA. Nine Filaments tapering, *fig.* 4. 5. the Antheræ fitting on the filaments, before the fhedding of the Pollen, oblong, reddifh, having four grooves, and terminated by a fhort point, *fig.* 6. 7. appearing afterwards fomewhat heart fhaped, flat, and as if compofed of two lamellæ, *fig.* 4. the Pollen is of a bright yellow colour.

PISTILLUM. Germen fubtriangulare, latere exteriore latiore, convexo, *fig.* 9. 10. Styli fex fubulati ; *fig.* 8. Stigma canaliculatum.

PISTILLUM. the Germen nearly triangular, the outer fide broad and roundifh, *fig.* 9. 10. fix Styles, tapering, the Stigma has a fmall channel in it, which afterwards fpreads into two lips, *fig.* 11. 12.

PERICARPIUM. Capsulæ fex, oblongæ, attenuatæ, erectæ, univalves, apice bilabiatæ, introrfum dehifcentes, *fig.* 11. 12.

SEED-VESSEL. Six Capsules, oblong, tapering, upright, of one valve, opening inwards, *fig.* 11. 12.

SEMINA plurima, minuta, oblonga, fufca, *fig.* 13.

SEEDS numerous, fmall, oblong, brown, *fig.* 13.

WE find this ftately Plant in, and by the fides of our watery ditches flowering from *July* to *September*. A few years fince it was found growing in *St. George's-Fields*, but the improvements making in that, and other parts adjacent to *London*, now oblige us to go farther in fearch of this and many other curious Plants. About the Ifland of *St. Helena*, near *Deptford*, and the Marfhes by *Blackwall*, it is found in great abundance, although very fcarce in many other parts of *Great-Britain*. Fifh-ponds, or other pieces of water, would derive great beauty from the introduction of this elegant native of our Ifle, the handfome appearance of which did not efcape our countryman, old Gerard, who defcribes it thus : " This Water Gladiole, or Graffy-Rufh, of all others, is the faireft and moft pleafant to behold, " and ferveth very well for the decking and trimming up of houfes, becaufe of the beautic and bravcric thereof."—— That accurate obferver Ray defcribes its nine Stamina, although in his time they were not viewed in that confequential light which they are in our prefent Syftems of Botany. It is the only Plant of the clafs Enneandria which grows wild in this kingdom. If vegetables were claffed according to their natural affinities, this would rank among the lillies. Cattle do not eat it. It is fo hardy as to bear the cold of *Lapland*.

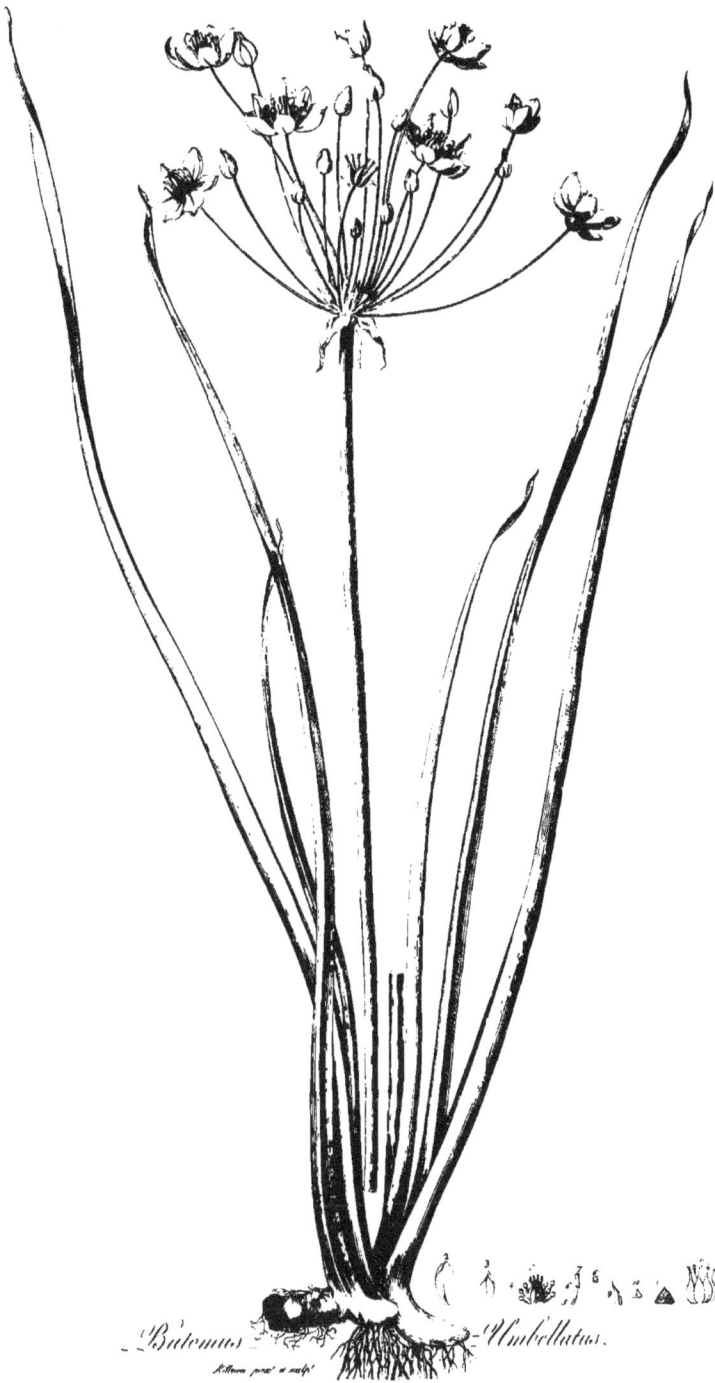

Butomus _Umbellatus._

Ritter pinx' et sculp'

Saxifraga granulata. White Saxifrage.

SAXIFRAGA *Linnæi Gen. Pl.* Decandria Digynia.

Calyx quinquepartitus. *Corolla* pentapetala. *Capfula* biroftris, unilocularis, polyfperma.

Raii Syn. Herbæ pentapetalæ vasculiferæ.

SAXIFRAGA *granulata* foliis caulinis reniformibus lobatis, caule ramofo, radice granulata. *Linn. Syft.*

Vegetab. p. 344. Fl. Suecic. n. 372.

SAXIFRAGA foliis radicalibus reniformibus, obtufe dentatis, caulinis palmatis. *Haller. hift. helv. n.* 976.

SAXIFRAGA rotundifolia alba. *Bauhin Pin.* 309.

SAXIFRAGA alba. *Gerard emac.* 841.

SAXIFRAGA alba vulgaris. *Parkinfon* 424. *Raii Syn.* 354. *Hudfon Fl. Angl. p.* 159. *Oeder. Flor. Dan.* 514.

RADIX. Fibris hujus radicis glomeratim adnafcuntur plurimi bulbilli, extus rubefcentes aut flavefcentes, intus albidi, faporis primum adftringentis, poftea amari et ingrati.

ROOT. To the fibres of the root of this plant, adhere in clufters a number of fmall bulbs, externally red or yellowifh, internally white, of a tafte at firft aftringent, afterwards bitter and difagreeable.

CAULIS plerumque fimplex, pedalis, fubramofus, teres, hirfutus, præfertim ad bafin, parum foliofus.

STALK generally fimple, about a foot high, a little branched, round, hirfute particularly at bottom, furnifhed with but few leaves.

FOLIA *radicalia* petiolis longis, hirfutis, bafi latis infidentia, reniformia, hirfutula, lobata, lobis obtufis ; *caulina* ficut adfcendunt petiolis brevioribus gaudent donec tandem feffilia fiunt, lobi foliorum acutiora evadunt, apicibus rufefcentibus.

LEAVES *which grows next the root* placed on long hairy foot-ftalks with a broad bafe, kidney-fhaped, flightly hairy, divided into obtufe lobes, thofe of the *ftalk*, as they afcend, are furnifhed with fhorter foot-ftalks, 'till they gradually become feffile, the lobes more acute, and the tips of a reddifh colour.

CALYX: Perianthium quinquepartitum, hirfutulum, fubvifcidum, laciniis ovato-acutis apice rufis, *fig.* 1.

CALYX : a Perianthium divided into five fegments, hirfute and fomewhat vifcid, the laciniæ of an oval pointed fhape, and red at the tips, *fig.* 1.

COROLLA : Petala quinque alba, patentia, apice rotundata, bafi anguftiora et venis flavefcentibus notata, *fig.* 2.

COROLLA : five Petals, white, fpreading, round at top, at bottom narrower, and ftriped with yellowifh veins, *fig.* 2.

STAMINA : Filamenta decem fubulata ; Antheræ ovatæ, compreffæ, infidentes, flavæ, biloculares, quorum quinque Polleu primum emittunt, hinc longiores, *fig.* 3, 4.

STAMINA : ten Filaments tapering ; Antheræ oval, flat, fitting on the Filaments, yellow, bilocular, five of them fhed the Pollen firft, hence they become longer than the others, *fig.* 3, 4.

PISTILLUM : Germen fubrotundum, inferum, glandula faturate viridi cinctum, *fig.* 7 ; Styli duo Staminibus breviores, incurvati, *fig.* 5 ; Stigma concavum, *fig.* 5, demum expandens, *fig.* 6.

PISTILLUM : Germen roundifh, placed below the Calyx, furrounded by a gland of a deep green colour, *fig.* 7 ; Styles, two, fhorter than the Stamina, bending inward, *fig.* 5 ; Stigma hollow, *fig.* 5, finally expanding, *fig.* 6.

PERICARPIUM : Capfula fubovata, biroftris, bilocularis, pallide fufca, *fig.* 8.

SEED-VESSEL : a Capfule of a fhape fomewhat oval, and pale brown colour, having two beaks or horns, and two cavities, *fig.* 8

SEMINA numerofa, minutiffima, nigra. *fig.* 9.

SEEDS numerous, very minute, and black. *fig.* 9.

THE Root of this fpecies of *Saxifrage*, by means of which it is chiefly propagated, affords the young Botanift a very good example of the *Radix granulata*, being compofed of a number of little grains or bulbs, connected together in clufters by the fibres ; fome of thefe bulbs are folid and entire, not unaptly refembling in fhape the bulbs of Onions ; others fpread open at top, and feem to be compofed of a number of fquamulæ or leffer bulbs, thefe are often of a bright red colour : the upper part of the ftalk, the foot-ftalks of the flowers, and calyx, are covered with a kind of hairs, which terminate in a vifcid globule, and which feem to accompany moft of the plants of this Genus. The two Styles, which at firft are fhort, with a hollow Stigma, *fig.* 5, quickly grow much longer ; the Stigmata fpread open, fo that they refemble in fome degree a pair of tea-tongs, *fig.* 6.

This plant does not occur fo frequently with us as many others ; according to Mr. Hudson, it is common about *Wandfworth* ; I have frequently gathered it in the fields about *Peckham*, and lately have found it in great abundance much nearer town, viz. in the fields called *Lock-fields*, on the right hand fide of *Kent-ftreet Road*, at the back of, and contiguous to Mr. Driver's Nurfery Gardens : it delights to grow in dry paftures which have a gravelly bottom ; flowers in May, and produces its feeds in the month following. When double, it ferves, with many other Britifh plants, to ornament the gardens of the curious.

Like many other plants, this feems to owe what little importance it has in medicine to the doctrine of fignatures, which has moft unphilofophically introduced a number of plants into our Materia Medica. As the root bore fo great a refemblance to little ftones, it was concluded it muft be efficacious in the ftone and gravel, for which difeafes it has been recommended, but there are no accounts of its fuccefs to be depended on. If it does poffefs any medical virtue, it fhould appear from the tafte of the root to be that of an aftringent.

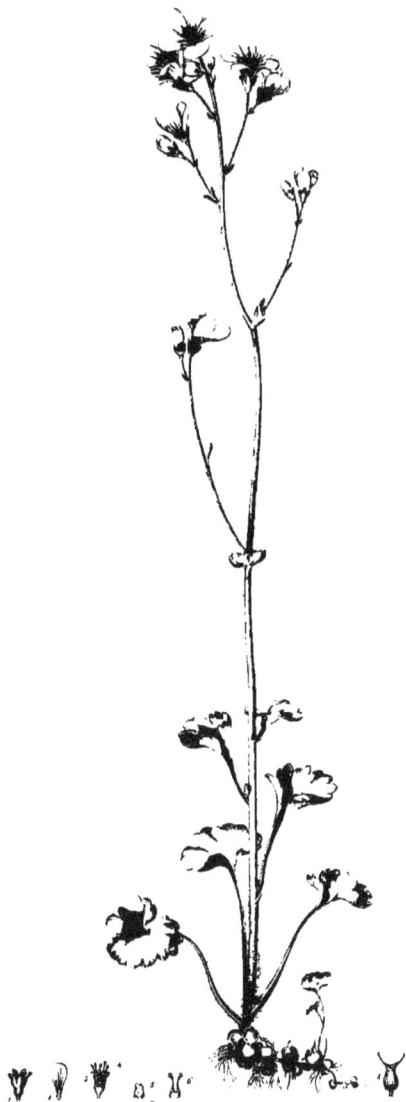

Saxifraga Granulata.

Sedum album. White-flowered Stonecrop.

SEDUM *Linnæi Gen. Pl.* Decandria Pentagynia.

Cal. 5 fidus. *Cor.* 5-petala. *Squamæ* nectariferæ 5, ad basin germinis. *Caps.* 5.

Raii Syn. Gen. 17. Herbæ multisiliquæ seu corniculatæ.

SEDUM *album* foliis oblongis obtusis teretiusculis sessilibus patentibus, cyma ramosa. *Linn. Syst. Vegetab. p.* 359. *Sp. Pl. p.* 619. *Fl. Suecic.* 153.

SEDUM caule glabro, foliis teretibus; umbellis ramosis; floribus petiolatis. *Haller hist. helv. n.* 959.

SEDUM *album. Scopoli Fl. Carn. p.* 324.

SEDUM minus teretifolium album. *Bauhin. p.* 283.

SEDUM minus officinarum. *Gerard emac.* 512.

VERMICULARIS five crassula minor vulgaris. *Parkinson* 734. *Raii Syn.* 271. *Hudson Fl. Angl. p.* 171. *Oeder. Fl. Dan. Icon.* 66.

RADIX perennis, fibrosa.

ROOT perennial and fibrous.

CAULES flexuose super muros repent, dein eriguntur, triunciales circiter, foliosi, rubri.

STALKS creep on the walls in a crooked form, then grow upright, about three inches high, leafy, and red.

FOLIA sessilia, oblonga et fere cylindracea, obtusa, non admodum conferta, patentia, carnosa, glabra, sæpius rubicunda.

LEAVES sessile, oblong and almost cylindrical, obtuse, but thinly placed on the stalk, spreading, fleshy, smooth, and generally of a reddish colour.

INFLORESCENTIA : Flores petiolati, in Cymam ramosam confertam dispositi.

INFLORESCENCE : Flowers standing on foot-stalks, and disposed in a thick branched Cyma.

CALYX : Perianthium pentaphyllum, foliolis brevibus, obtusis, *fig.* 1.

CALYX : a Perianthium of five leaves, which are short and obtuse, *fig.* 1.

COROLLA : Petala quinque alba, acuminata, lineâ longitudinali rubra sæpius notata, *fig.* 2.

COROLLA : five white Petals, acuminated and generally marked with a longitudinal red streak, *fig.* 2.

NECTARIUM glandula minima squamiformis ad basin singuli Germinis. *fig.* 6.

NECTARY a very minute squamiform gland at the base of each of the Germina, *fig.* 6.

STAMINA : Filamenta decem alba, *fig.* 2, 3; Antheræ rubræ.

STAMINA : ten white Filaments, *fig.* 2, 3; Antheræ deep red.

PISTILLUM : Germina quinque, in Stylos totidem acuminatos desinentia ; Stigmata simplicia, *fig.* 4, 5.

PISTILLUM : five Germina, terminating in so many acuminated Styles ; the Stigmata simple, *fig.* 4, 5.

PERICARPIUM : Capsulæ quinque minimæ acuminatæ introrsum dehiscentes, *fig.* 7.

SEED-VESSEL : five small acuminated Capsules, opening inwardly, *fig.* 7.

SEMINA parva, oblonga, *fig.* 8.

SEEDS small and oblong, *fig.* 8.

The *Sedum album* may be considered with us as rather a scarce plant ; it is found here and there on the Walls about Town, particularly on the Chapel-wall in *Kentish-Town*, where it has grown for many years; also upon a Wall on the left-hand side leading from *Bromley* to *Bromley-Hall*, in *Middlesex*. It has been thought to possess sufficient beauty to recommend it as a garden plant, and is accordingly, with very little trouble, cultivated in many of the gardens of the curious, nothing more being necessary than placing it in a pot filled with gravel or mould : in such a situation it will grow, flourish, and propagate itself very fast.

It has been called *album* from the colour of its flowers, which generally however have a tinge of red in them. It flowers in July. The round and oblong shape of its leaves readily distinguishes it from our other *Stonecrops*.

Haller informs us that it possesses all the virtues of the large *Houseleek*, and that he has used the juice of it in uterine hæmorrhages, but does not inform us with what success. By way of cataplasm it is applied to the piles when in a painful state, and is said to have sometimes been made the same use of in cancers with success. By some it is eaten as a pickle.

Sedum Album.

SEDUM ACRE. COMMON YELLOW STONECROP, OR WALL-PEPPER.

SEDUM. *Linnæi Gen. Pl.* DECANDRIA PENTAGYNIA.

Raii Synopsis Gen. 17. HERBÆ MULTISILIQUÆ SEU CORNICULATÆ.

SEDUM *acre* foliis fubovatis, adnato-feffilibus, gibbis, erectiufculis, alternis; cyma trifida. *Lin. Syst. Vegetab.* p. 359. *Fl. Suecic.* p. 153.

SEDUM foliis conicis confertis, caulibus ramofis, fummis trifidis. *Haller. hist. v.* 1. *n.* 966.

SEMPERVIVUM minus vermiculatum acre. *Bauhin. pin.* 283.

VERMICULARIS feu Illecebra minor acris. *Ger. emac.* 517.

ILLECEBRA minor feu Sedum tertium Diofcoridis. *Parkinfon* 735. *Raii Synop.* 270. *Hudfon. Fl. Angl.* p. 171.

RADIX perennis, fibrofa.	ROOT perennial, and fibrous.
CAULES numerofi, cæfpitofi, ramofiffimi, palmares, ad bafin repentes, dein erecti, teretes, foliofiffimi.	STALKS numerous, growing in tufts, very much branched, three inches high, creeping at their bafe, but afterwards growing upright, round, and very leafy.
FOLIA alterna, conferta, imbricata, fuberecta, adnato-feffilia, ovata, obtufa, brevia, carnofa, margine paululùm comprefta, glabra, *fapore acri. fig.* 1.	LEAVES alternate, growing very thick together, and laying one over another, nearly upright, growing to the ftalk, oval, blunt, fhort, flefhy, flattened a little at the edges, fmooth, and of a *very biting tafte*, fig. 1.
FLORES feffiles, lutei, in Cymas fubtrifidas difpofiti.	FLOWERS feffile, yellow, growing in Cymæ fomewhat trifid.
CALYX: PERIANTHIUM quinquepartitum, perfiftens, laciniis craffis obtufiufculis, *fig.* 2.	CALYX: a PERIANTHIUM divided into five fegments, and continuing, the fegments thick and bluntifh, *fig.* 2.
COROLLA: PETALA quinque lanceolato-acuminata, plana, patentia, Calyce duplo longiora, *fig.* 3.	COROLLA: compofed of five long-pointed PETALS which are flat, fpreading, and twice the length of the Calyx, *fig.* 3.
NECTARIUM: Squamula minima, alba, ad bafin, finguli germinis extrorfum pofita, *fig.* 7.	NECTARY: a very minute fcale or gland placed externally at the bottom of each Germen, *fig.* 7.
STAMINA: FILAMENTA decem fubulata, longitudine Corollæ. ANTHERÆ flavæ, *fig.* 4.	STAMINA: ten FILAMENTS, tapering, the length of the Corolla, the ANTHERÆ yellow, *fig.* 4.
PISTILLUM: GERMINA quinque oblonga, flava, in STYLOS acuminatos definentia. STIGMATA fimplicia, *fig.* 6.	PISTILLUM: five GERMINA, oblong, yellow, terminating in five long-pointed STYLES. The STIGMATA fimple, *fig.* 6.
PERICARPIUM: CAPSULÆ quinque patentes, acuminatæ, compreffæ, longitudinaliter futura introrfum dehifcentes, *fig.* 8.	SEED-VESSEL: five CAPSULES, fpreading, long-pointed, flat, opening internally by a longitudinal future, *fig.* 8.
SEMINA minima, ovata, rufa, *fig.* 9.	SEEDS very minute, oval, and reddifh brown, *fig.* 9.

According to the account which fome medical Writers give of this Plant it appears to poffefs confiderable virtues, while others, from the durability of its acrimony, and the violence of its operation, have thought it fcarce fafe to be adminiftered. Chewed in the mouth it has a very hot and biting tafte, whence its name of *Wall-Pepper*. Applied to the fkin it excoriates and exulcerates it, taken internally it proves emetic and diuretic.

The Difeafes in which it has been chiefly recommended are the Scurvy and Dropfy, in both of which, according to Linnæus, it is an excellent remedy; and fome inftances are brought of the efficacy of its juice in Cancers, but thefe perhaps, ftand in need of farther confirmation.

It grows very common on Houfes, Walls, and gravelly Banks, and flowers in June; it continues but a fhort time in bloffom, but while it lafts its lively yellow colour gives a very pretty appearance to thofe Houfes and Walls which are covered with it.

(Sedum acre?)

Lychnis Flos cuculi.

LYCHNIS FLOS CUCULI. MEADOW LYCHNIS.

LYCHNIS *Linnæi Gen. Pl.* Decandria Pentagynia.

Raii Synopſis Gen. 24. Herbæ pentapetalæ vasculiferæ.

LYCHNIS *Flos Cuculi petalis quadrifidis fructu ſubrotundo. Lin. Syſt. Vegetab. p.* 361. *Sp. Pl.* 625.

LYCHNIS petalis quadrifidis. *Haller. hiſt. v.* 1. *n.* 921.

CARYOPYLLUS pratenſis, laciniato flore ſimplici, five Flos cuculi. *Bauhin. pin.* 210.

LYCHNIS plumaria ſylveſtris ſimplex. *Parkinſon. parad.* 253.

ARMERIUS pratenſis mas et fœmina. *Gerard. Emac.* 600.

Raii Synop. ed. 3. 338. *Hudſon. Fl. Angl.* 174. *Order. Flor. Dan. tab.* 590. *Scopoli. Fl. Carniol. ed.* 2. *p.* 311.

RADIX perennis, fibroſa, ex albido fuſca, ſaporis ſub-acris.	ROOT perennial, fibrous, of a browniſh white colour, and ſomewhat biting taſte.
CAULIS pedalis ad tripedalem, erectus, ſulcato-angulatus, articulatus, geniculi tumidi, ſcabriuſculus, purpuraſcens.	STALK from one to three feet high, upright, ſomewhat angular and grooved, jointed, the joints ſwelled, roughiſh, and of a purpliſh colour.
FOLIA Caulis, oppoſita, connata, lanceolata, carinata, ſuberecta, lævia.	LEAVES of the Stalk oppoſite, connate, lanceolate, the midrib prominent underneath, upright and ſmooth.
PEDUNCULI oppoſiti, plerumque unico intermedio.	PEDUNCLES oppoſite, one generally intermediate.
CALYX: Perianthium monophyllum, tubulatum quinquedentatum, decangulatum, purpureum, perſiſtens. *fig.* 1.	CALYX a Perianthium of one leaf, tubular, quinquedentate, having ten angles, or ridges, and of a deep purple colour.
COROLLA Petala quinque, unguis longitudine calycis, *fig.* 2. Limbus quadrifidus, laciniis exterioribus brevioribus, *fig.* 4. ad baſin limbi laminæ duæ erectæ acutæ. *fig.* 3.	COROLLA five petals, the claw the length of the Calyx, *fig.* 2. the Limb divided into four laciniæ, the exterior ſhorteſt and narroweſt, *fig.* 4. at the bottom of the limb are placed two ſmall upright laminæ, *fig.* 3.
STAMINA: Filamenta decem, ſubulata, quorum quinque breviora, *fig.* 5, brevioribus ungui petalorum affixis. *fig.* 6. Antheræ oblongæ, biloculares, *fig.* 7. incumbentes, purpuraſcentes.	STAMINA: ten Filaments, tapering, five long and five ſhort, *fig.* 5. the ſhorter filaments affixed to the claw of each petal, *fig.* 6. the Antheræ oblong, bilocular. *fig.* 7. laying acroſs the filaments, and of a purpliſh hue.
PISTILLUM Germen ſubovatum, *fig.* 8. Styli quinque ſubulati, ſubincurvati, *fig.* 10. Stigmata ſimplicia. *fig.* 10.	PISTILLUM: the Germen ſomewhat oval, *fig.* 8. five Styles tapering and bending a little inward, *fig.* 10. Stigmata ſimple. *fig.* 10.
PERICARPIUM Capsula ovata, unilocularis, ore quinquedentato, dentibus reflexis. *fig.* 9.	SEED-VESSEL: a Capsule, oval, of one cavity, the mouth having five teeth which turn back. *fig.* 9.
SEMINA numeroſa, ſubcompreſſa, ſcabriuſcula, ex cinereo-fuſca. *fig.* 11. 12.	SEEDS numerous, flattiſh, rough, and of a brown aſh colour. *fig.* 11. 12.

A variety of names hath been given to this Plant, as Meadow Pink, Cuckow Flower, Wild Williams, Ragged Robin, &c. Meadow Lychnis however ſeems to us the moſt eligible. It abounds in moiſt Meadows, where it flowers in May and June, and is included amongſt the great number of which our Meadow hay is compounded. Goats, Sheep, and Horſes are ſaid to feed on it. The uſe to which it is applied, ſeems to be chiefly ornamental; the beauty of its flowers juſtly entitles it (with many other neglected Britiſh Plants) to a place in the Gardens of the curious: where it is frequently found with a double flower, making a beautiful appearance, and requiring little more care in its culture, than to be placed in a moiſt ſituation: It may be propagated either by ſeeds or ſlips; the ſeeds may be found ripe in the latter end of June, by the ſides of ditches, where the Mower's Scythe has not reached them. We ſometimes find the Meadow Lychnis growing wild with a double flower, and ſometimes with a white one; but this is altogether accidental.

The agreement between the blowing of flowers, and the periodical return of birds of paſſage, has been attended to from the earlieſt ages: Before the return of the ſeaſons was exactly aſcertained by Aſtronomy, theſe obſervations were of great conſequence in pointing out ſtated times for the purpoſes of Agriculture; and ſtill, in many a Cottage, the birds of paſſage and their correſponding flowers aſſiſt in regulating

"The ſhort, and ſimple Annals of the Poor."

For this reaſon, no doubt, we have ſeveral other plants that, in different places, go by the name of Cuckow Flower. *Gerard* ſays, Cardamine pratenſis (Common Ladies Smock) is the true Cuckow Flower. *Shakeſpear's* Cuckow Buds are or "yellow hue." By ſome the Orchis, Arum, and Wood-ſorrel are all called after the Cuckow.

Cerastium aquaticum.

CERASTIUM AQUATICUM. MARSH CERASTIUM OR MOUSE-EAR CHICKWEED.

CERASTIUM *Linnæi Gen. Pl.* DECANDRIA PENTAGYNIA.

Raii *Synop. Gen:* 24 HERBÆ PENTAPETALÆ VASCULIFERÆ.

CERASTIUM *aquaticum foliis cordatis, feſſilibus, floribus folitariis, fructibus pendulis. Linnæi Syſt: Vegetab. p. 363. Fl. Suecic. p.* 157.

ALSINE foliis ovato-cordatis, imis petiolatis, tubis quinis. *Haller. hiſt. n.* 885.

STELLARIA aquatica. *Scopoli Fl. Carniol. p.* 320.

ALSINE aquatica major. *Bauhin. pin.* 254.

ALSINE major. *Gerard emac.* 611. maxima *Parkinſon* 759. *Raii Syn. p.* 347. *Hudſon Fl. Angl. p.* 177.

RADIX perennis, fibroſa, repens.

ROOT perennial, fibrous, and creeping.

CAULES bipedales, debiles, pene teretes, teneri, filoſi, hirſuti, ramoſi, rami alterni.

STALKS about two feet in length, weak, almoſt round, tender, ſtringy, hirſute, and branched, the branches alternate.

FOLIA Caulis feſſilia, amplexicaulia, cordato-acuminata, margine in ſuperioribus preſertim undulata, lævia, ſubviſcida; ramorum magis undulata, petiolata.

LEAVES of the Stalk feſſile, embracing the Stalk, ſomewhat heart ſhaped and acuminate, the edge particularly in the upper ones waved, ſmooth, and ſomewhat viſcid; thoſe of the branches more waved with ſhort footſtalks.

PEDUNCULI alterni, e dichotomia caulis, uniflori, *poſt floreſcentiam penduli.*

FOOTSTALKS alternate, from the forking of the Stalk, uniflorous, *after the bloſſom is gone off pendulous.*

CALYX: PERIANTHIUM pentaphyllum, perſiſtens, foliolis lanceolatis, concavis, ſubcarinatis, apice obtuſiuſculis, hirſutis, margine membranaceis, petalis paulo brevioribus. *fig.* 1.

CALYX: a PRIANTHIUM of five leaves, perſiſting, the leaves lanceolate, concave, ſlightly keel-ſhaped, bluniſh at top, hirſute, at the edge membranous, and a little ſhorter than the Petals, *fig.* 1.

COROLLA: PETALA quinque alba, patentia, bipartita, laciniis oblongis, nervoſis, divaricantes, *fig.* 2. 3.

COROLLA: five PETALS white, ſpreading, divided almoſt to the bottom, the laciniæ or ſegments oblong, nervous, and divaricating, *fig.* 2. 3.

STAMINA: FILAMENTA decem, ſubulata, alba, receptaculo inſerta, ad baſin et inter petala alterne locata, *fig.* 4. quæ inter petala locantur paulo longiora ſunt et glandula ad baſin inſtruuntur *fig.* 5. ANTHERÆ inſidentes, biloculares, albæ, *fig.* 4.

STAMINA: ten FILAMENTS, tapering, white, fixed to the receptacle, placed alternately, one at the baſe and one betwixt each petal, *fig.* 4; thoſe placed between the petals are a little longer than the others, and furniſhed at bottom with a gland, *fig.* 5. ANTHERÆ white and bilocular, *fig.* 4.

PISTILLUM: GERMEN ſubrotundum, apice ſulcatum, STYLI quinque albi, filiformes, longitudine Germinis. STIGMATA ſimplicia, *fig.* 6.

PISTILLUM: GERMEN roundiſh, at top grooved, five STYLES thread-ſhaped, white, the length of the Germen. STIGMATA ſimple, *fig.* 6.

PERICARPIUM: CAPSULA ovata, obſolete pentagona, ore quinquedentato. *fig.* 7.

SEED-VESSEL: an oval CAPSULE, ſlightly pentangular, the mouth quinquedentate.

SEMINA rufa, ſubreniformia, ſcabra, 60 numeravi, *fig.* 8. 9.

SEEDS reddiſh brown, rough, about 60 in each capſule, *fig.* 8. 9.

SOME of our modern and moſt celebrated ſyſtematic Botaniſts ſeem very much divided with reſpect to the Genus to which this Plant ſhould belong. HALLER makes it an *Alſine* or *Chickweed*; SCOPOLI a *Stellaria*, and LINNÆUS a *Ceraſtium*. We ſhall not pretend to decide who is moſt in the right, but only obſerve that its general habit or appearance, and the form of its ſeeds, might eaſily induce HALLER to conſider it as an *Alſine*; the ſhape of its petals, with the ſtructure of its ſeeds, would juſtify SCOPOLI in calling it a *Stellaria*, while the number of its ſtyles might lead LINNÆUS with propriety to place it among the *Ceraſtiums*. To us it appears to have the greateſt natural affinity with the *Alſine media* or common *Chickweed*; it is true LINNÆUS ranks that plant among thoſe which have *five* Stamina, yet it is frequently obſerved to have more, and the ſtructure of the flower evidently ſhows it to be formed for having *ten*, and thoſe flowers which have not that number may be conſidered as imperfect. The Seeds of theſe two plants are ſo ſimilar as ſcarcely to be diſtinguiſhed from each other, and their ſtalks are procumbent, tender, brittle, and ſtringy, indeed they frequently ſo much reſemble one another, as to oblige the young Botaniſt to have recourſe to the very different ſize of their flowers in order to diſcriminate them.

This Plant grows in moiſt places, on the banks of rivers and by ſtreams of water, it flowers in July and Auguſt. SCOPOLI aſſerts that the plants of this kind afford excellent food for Kine.

Euphorbia Peplus. Small Garden Spurge.

EUPHORBIA *Linnæi Gen. Pl.* Dodecandria Trigynia.
  Raii Syn. Gen. 22. Herbæ vasculiferæ flore tetrapetalo anomalæ.
EUPHORBIA *(Peplus)* umbella trifida, dichotoma, involucellis ovatis, foliis integerrimis obovatis petiolatis.
  Linn. Syst. Vegetab. p. 375. *Fl. Suecic. p.* 163.
TITHYMALUS foliis rotundis, ſtipulis floralibus cordatis, obtuſis, petalis argute corniculatis. *Haller. hiſt.*
  vol. 2. *f.* 9. *n.* 1049.
PEPLUS five Eſula rotunda. *Bauhin pin.* 292. *Parkinſon. Gerard. emac.* 503.
TITHYMALUS parvus annuus, foliis ſubrotundis non crenatis, Peplus diĉtus. *Raii Syn. p.* 313. *n.* 9:
  Petty Spurge. *Hudſon Fl. Angl. p.* 182.

RADIX annua, lignoſa, ſimplex, fibroſa, albida.	ROOT annual, woody, ſimple, fibrous and whitiſh.
CAULIS, ſubereĉtus, dodrantalis, teres, glaber, ramoſus, baſi durior, tenuior, ſubruber, folioſus, laĉtiſuus.	STALK generally upright, about nine inches high, round, ſmooth, and branched; at bottom harder, more ſlender, and of a reddiſh colour, leafy and milky.
RAMI pauci, ſparſi, inferioribus longioribus oppoſitis.	BRANCHES few, not growing in any regular order, the lower ones longeſt and oppoſite.
UMBELLA triſida, dichotoma.	UMBEL firſt trifid, then dichotomous.
FOLIA obovata, petiolata, integerrima, ſparſa, obtuſiuscula, inferioribus ſubrotundis.	LEAVES ſomewhat oval, but narroweſt towards the baſe, having foot-ſtalks, entire at the edges, placed in no regular order, ſomewhat blunt, the lowermoſt leaves almoſt round.
STIPULÆ umbellæ tres, ovato-acutæ, petioſis brevibus inſidentes, umbellulæ alterne oppoſitæ, ſeſſiles, cordato-ovatæ, inæquales, integerrimæ, baſi quâ tendit germen quaſi excavatæ.	STIPULÆ of the large umbel three in number, oval and pointed, placed on very ſhort foot-ſtalks: of the ſmall umbel alternately oppoſite, ſeſſile, of an heart-ſhaped-oval form, unequal, and entire, at bottom on that ſide to which the Germen tends as if cut away.
CALYX ventricoſus, perſiſtens. fig. 1.	CALYX bellying out and continuing, fig. 1.
COROLLA nulla.	COROLLA wanting.
NECTARIA quatuor bicorniculata, fig. 2.	NECTARIES four, each having two little horns, fig. 2:
STAMINA plerunque duo, aut tria, viſibilia, exſerta: Antheræ didymæ, ſubrotundæ, fig. 3.	STAMINA ſeldom more than two or three, which are viſible, and placed without the Calyx: Antheræ two on each filament joined together, of a roundiſh figure, fig. 3.
PISTILLUM: Germen pedunculatum, nutans, triangulare, angulis longitudinaliter ſulcatis, fig. 4, 6: Stigmata tria, apice biſida, fig. 5.	PISTILLUM: Germen placed on a foot-ſtalk, hanging down, triangular, the angles longitudinally grooved, fig. 4, 6: Stigmata three, biſid at top, fig. 5.
PERICARPIUM: Capſula tricocca, trilocularis, trivalvis, valvulis lævibus, et dum adhuc virides diſſilientibus, fig. 6.	SEED-VESSEL: a Capſule of three cavities, and three valves, the valves protuberant, ſmooth, and ſplitting with a kind of elaſticity even while they are of a green colour, fig. 6.
SEMEN unicum in ſingulo loculamento, ovatum, canum, elevatatum, appendiculatum, fig. 7.	SEED one in each cavity, oval, grey, with numerous depreſſions on its ſurface, and a little white button at one end, fig. 7.
N. B. Omnes partes fructificationis lente augentur.	N. B. All the parts of fructification are magnified.

MANY of the Spurges conſiderably reſemble one another, and two of them that have this affinity, grow frequently together in Gardens, viz. the preſent Spurge, *Euphorbia Peplus*, and the Sun Spurge, *Euphorbia Helioſcopia*; they may be diſtinguiſhed however by the ſlighteſt attention. In the *Helioſcopia* the leaves are notched or ſerrated at the edges, in the *Peplus* they are entire, in the *Helioſcopia* the Petals or rather Neĉtaria are round and entire, in the *Peplus* each is furniſhed with two little horns, fig. 2; there are other marks of diſtinĉtion but theſe are the moſt ſtriking. This ſpecies grows in Gardens and other cultivated ground, and flowers in Autumn.

The milky fluid which it abounds with, is by ſome applied to Warts, which it is ſaid to deſtroy.

Moſt if not all the plants of this Genus contain in them this milky and gummy ſubſtance, which to the taſte is exceedingly acrid; and this laĉtiſuous property, joined to the peculiarity of its parts of fructification, point out almoſt at firſt ſight this natural family of plants. But the botanic Student who would inveſtigate this ſpecies according to the principles of the Linnæan Syſtem, not having theſe charaĉteriſtics to aſſiſt him, finds a conſiderable difficulty in learning even the *Claſs* to which it belongs, nor is it poſſible for him to aſcertain the Claſs by an examination of this or ſcarce any other Engliſh Spurge: the *Stamina* in the firſt place are very minute, it is ſeldom that more than two or three protrude beyond the Calyx, all the reſt lye concealed within it, they ſeldom amount to twelve in number, and even if they did amount to that exaĉt number, their minuteneſs and the milky juice which flows from the diſſection, render the enumeration of them ſcarce praĉticable. The Student may however in a great degree ſurmount this difficulty, by an examination of ſome plant of this genus, which is larger in every reſpeĉt, and the *Euphorbia Lathyris* improperly called the *Caper Tree*, (which is cultivated in many Gardens) will afford him a very good example, and tend to give him a clear idea of the flower and fruit of this ſingular genus of plants.

I would not be thought on account of this difficulty to inveigh againſt Linnæus's Syſtem, being ſenſible that difficulties occur, and muſt occur in all botanic arrangements, and inſtead of ſelecting faults inſeparable from every mode of claſſification, (which ſeems to have been a favourite amuſement of ſome Authors, and forms indeed the greateſt part of their writings) I would uſe every endeavour to make it more perfeĉt.

It is too much the faſhion now, as well as formerly, for every Botaniſt as ſoon as he thinks he has ſome pretenſions to eminence, to ſet about the arduous taſk of framing a new Syſtem; he may by this means give the public ſome idea of his ſelf-conſequence, and be inrolled in the Catalogue of Syſtem-makers, but not one jot will he advance the ſcience of Botany. It is to be regretted that Botaniſts will not be contented with a Syſtem, a proof of whoſe ſuperiority is the almoſt general reception it has met with throughout Europe, and unite in their endeavours to render that Syſtem more compleat, by giving us an accurate account of the hiſtory of thoſe plants not already given, their virtues and uſes; this appears to me to be the true method of advancing this delightful Science, and making it uſefull to Mankind.

When one Syſtem of Botany is generally followed as is nearly the caſe at preſent, Botaniſts in different kingdoms perfeĉtly underſtand each others language, but when each adopts a ſeperate one, (which is frequently diĉtated by Pride or Caprice) all becomes Babel; and every one who wiſhes to acquire a knowledge of the plants treated of, muſt at conſiderable expence both of time and labour, acquire firſt the Authors new-created Syſtem-language, a tax which it is hoped every true Botaniſt will unite to oppoſe.

Euphorbia Peplus

EUPHORBIA HELIOSCOPIA. SUN SPURGE OR WART-WORT.

EUPHORBIA *Linnæi Gen. Pl.* DODECANDRIA TRIGYNIA.

Cor. 4-f. 5-petala, calyci infidens. *Cal.* 1-phyllus, ventricofus.

Capf. 3-cocca.

Raii Syn. Gen. 22. HERBÆ VASCULIFERÆ, FLORE TETRAPETALO ANOMALÆ.

EUPHORBIA umbella quinquefida : trifida : dichotoma, involucellis obovatis, foliis cuneiformibus ferra-tis. *Linn. Syft. Vegetab. p.* 377. *Sp. Plant.* 658. *Fl. Suecic. p.* 162.

TITHYMALUS foliis petiolatis, fubrotundis, ferratis, ftipulis rotundis, ferratis. *Haller hift. v.* 2. *p.* 10. *n.* 1050.

TITHYMALUS *heliofcopius. Scopoli Fl. Carniol. p.* 337. *n.* 579.

TITHYMALUS *heliofcopius. Bauhin Pin.* 291. *Gerard emac.* 458. *Parkinfon.* 189.

TITHYMALUS *heliofcopius* five folifequus. *I. B.* 3. 669. *Raii Syn.* 313. *Hudfon Fl. Angl. p.* 183.

RADIX fimplex, fibrofa, annua.

ROOT fimple, fibrous, annual.

CAULIS erectus, teres, pilofus, inferne brachiatus, brachiis oppofitis.

STALK upright, round, flightly hairy, below branch-ed, the branches oppofite.

FOLIA fparfa, pauca, glabra, *ferrata*, cuneiformia, in-feriora petiolata, fuperiora feffilia.

LEAVES growing in no regular order, few, fmooth, *ferrated*, and wedge-fhaped, the lower ones ftanding on foot-ftalks, the upper ones feffile.

UMBELLA quinquefida, trifida, dichotoma, patens, faftigiata.

UMBELL dividing into five, next three, then two, fpreading, of an equal height at top.

STIPULÆ minute ferratæ, glabræ, UMBELLÆ quinque, obovatæ, horizont les, æquales, *Umbellulæ* tres, ovatæ, inæquales, interiore duplo minore, quæ fequuntur mucrone terminatæ.

STIPULÆ minutely ferrated and fmooth, thofe of the UMBELL five, fomewhat oval, fpreading hori-zontally, and equal ; thofe of the *fmaller* UM-BELL three, oval, unequal, the interior one twice as fmall as the others; thofe which follow terminating in a point.

CALYX fubventricofus, flavefcens, *fig.* 1.

CALYX fomewhat fwelled, of a yellowifh colour, *fig.* 1.

COROLLA nulla.

COROLLA wanting.

NECTARIA quatuor, fubrotunda, nuda, *fig.* 2.

NECTARIA four, roundifh and naked, *fig.* 2.

STAMINA : FILAMENTA duo, tria, aut plura, vifi-bilia, exferta ; ANTHERÆ flavæ, biloculares, loculis fubrotundis, *fig.* 3.

STAMINA : two, three, or more FILAMENTS, vifible beyond the Calyx ; ANTHERÆ yellow, bilocu-lar, the cavities containing the Pollen roundifh, *fig.* 3.

PISTILLUM: GERMEN pedunculatum, fubrotundum, nutans; STIGMATA tria, apice bifida, *fig.* 4, 5.

PISTILLUM: GERMEN placed on a foot-ftalk, round-ifh, hanging down; STIGMATA three, bifid at top, *fig.* 4, 5.

PERICARPIUM: CAPSULA tricocca, trilocularis, tri-valvis, *fig.* 6.

SEED-VESSEL a CAPSULE of three protuberating valves, and three cavities, *fig.* 6.

SEMEN unicum in fingulo loculamento, ovatum, rugo-fum ex purpureo fufcum, *fig.* 7.

SEEDS one in each cavity, oval, wrinkled, of a purp-lifh brown colour, *fig.* 7.

IN fpeaking of the *Euphorbia Peplus*, I had occafion to take notice of the difficulty which Students in Bo-tany find in invefligating the *Clafs* and *Order* of this Genus, and endeavoured to make it eafier to them: in this plant the parts of the fructification are fomewhat larger ; and it differs from the other Spurges in having its leaves finely ferrated. In its acrimonious quality it is inferior to none ; hence it has often been applied to Warts for the purpofe of deftroying them ; but even in this cafe, great care fhould be ufed in its applica-tion. My friend Mr. WILLIAM WAVELL lately informed me of a cafe which fell under his notice in the Ifle of Wight, where from the application of the juice of this Spurge to fome Warts near the eye of a little girl, the whole face became inflamed to a very great degree.

It is very common in gardens and cultivated ground, flowering in Autumn.

Euphórbia Heliòscopia

POTENTILLA REPTANS. COMMON CINQUEFOIL OR

FIVE LEAVED GRASS.

POTENTILLA *Linnæi Gen. Pl.* ICOSANDRIA POLYGYNIA:

Raii Gen. 15. HERBÆ SEMINE NUDO POLYSPERMÆ.

POTENTILLA *reptans* foliis quinatis, caule repente, pedunculis unifloris. *Lin: Syst. Vegetab. p.* 398. *Fl. Suecic. p.* 178.

FRAGARIA foliis quinatis ferratis, petiolis unifloris, caule reptante. *Haller hist. v.* 2. *p.* 47.

QUINQUEFOLIUM majus repens. *Bauhin pin. p.* 325. *Gerard emac.* 987.

PENTAPHYLLUM vulgatissimum *Parkinson* 398. *Raii Syn. p.* 255.

POTENTILLA *reptans. Hudson. Fl. Angl. p.* 197. *Scopoli Fl. Carniol. p.* 361

RADIX perennis, fusiformis, paucis fibrillis instructa, intra terram profunde penetrans, crassitie digiti minimi aut pollicis etiam in annosis, externe sordide castanea.

ROOT perennial, tapering, furnished with few fibres, penetrating deeply into the earth, the size of the little finger, or even of the thumb when old, externally of a dark chesnut colour.

CAULES numerosi, teretes, glabri, repentes, purpurei.

STALKS numerous, round, smooth, and creeping.

FOLIA quinata, etiam septena occurrunt, ferrata, venosa, inæqualia, parum hirfuta, petiolis longis initidentia, per paria e geniculis caulium ad magna intervalla prodeuntia.

LEAVES quinate, or growing five together, sometimes even seven, serrated, veiny, unequal in their size, slightly hairy, sitting on long footstalks, which proceed in pairs from the joints of the stalks at considerable distances.

STIPULÆ geminæ, trifoliatæ, foliolis ovatis.

STIPULÆ growing in pairs, composed of three oval-shaped leaves.

PETIOLI uniflori, longi, suberecti.

FOOT-STALKS of the flowers uniflorous, long, and nearly upright.

CALYX: PERIANTHIUM monophyllum, planiusculum, decemfidum, laciniis alternis minoribus, sæpe reflexis, *fig.* 3, 4, 5.

CALYX: a PERIANTHIUM of one leaf, flattish, divided into ten segments, the segments alternately smaller and frequently turned back, *fig.* 3. 4. 5.

COROLLA: PETALA quinque, fubrotundo-cordata, flava, unguibus calyci inferta, *fig.* 6.

COROLLA: five PETALS of a roundish heart-shaped figure, and yellow colour, inserted into the Calyx by their Ungues or claws, *fig.* 6.

STAMINA: FILAMENTA viginti, fubulata, Corolla breviora, margini interiori glandulofæ calycis inferta, in duas feries distributa; ANTHERÆ oblongæ, compressæ, flavæ, biloculares, loculæ membrana divisæ, infidentes, *fig.* 7, 8.

STAMINA twenty FILAMENTS tapering: shorter than the Corolla, inserted into the inner edge of the Calyx, which puts on a glandular appearance, and placed in two rows; ANTHERÆ oblong, flat, bilocular, the bags or cavities divided by a membrane, fitting on the filaments, *fig.* 7, 8.

PISTILLUM: GERMINA numerofa, in capitulum collecta; STYLI filiformes filamentis breviores, lateri Germinis inferti, perfiftentes; STIGMATA minima, obtuſa, *fig.* 9, 10.

PISTILLUM: the GERMINA numerous, collected into a little head; the STYLES filiform, shorter than the filaments, inserted into the fide of the Germen and continuing; the STIGMATA very small and blunt, *fig.* 9, 10.

SEMINA numerofa, parva, fufca, ſtylo perfiftente terminata, *fig.* 11, 12.

SEEDS numerous, small, brown and terminated by the Style, *fig.* 11, 12.

The Roots of *Cinquefoil* and many other plants of the Class *Icofandria*, poffefs confiderable virtues as aftringent medicines, and may be ufed in the fame Cafes in which *Biftort* is recommended.

It has likewife been ufed in fome places for the purpofe of tanning Leather where better materials for that purpofe are with difficulty acquired.

A Tea or infufion of the leaves is in ufe among the Country People as a drink in Fevers.

Moſt forts of Cattle are fond of the leaves, but it does not appear to be a plant worth cultivating on that account. The Larva or Caterpillar of the *Phalæna Rubi, vid. Roefel, Suppl. tab.* 69, *Albin tab.* 81, feeds on the leaves in Autumn, although a plant to which that Infect is by no means confined.

It grows very common in meadows and on banks by the road fides, and flowers in July, Auguft, and September. It affords the botanic Student a very good example of the *Caulis repens* or *Creeping Stalk.*

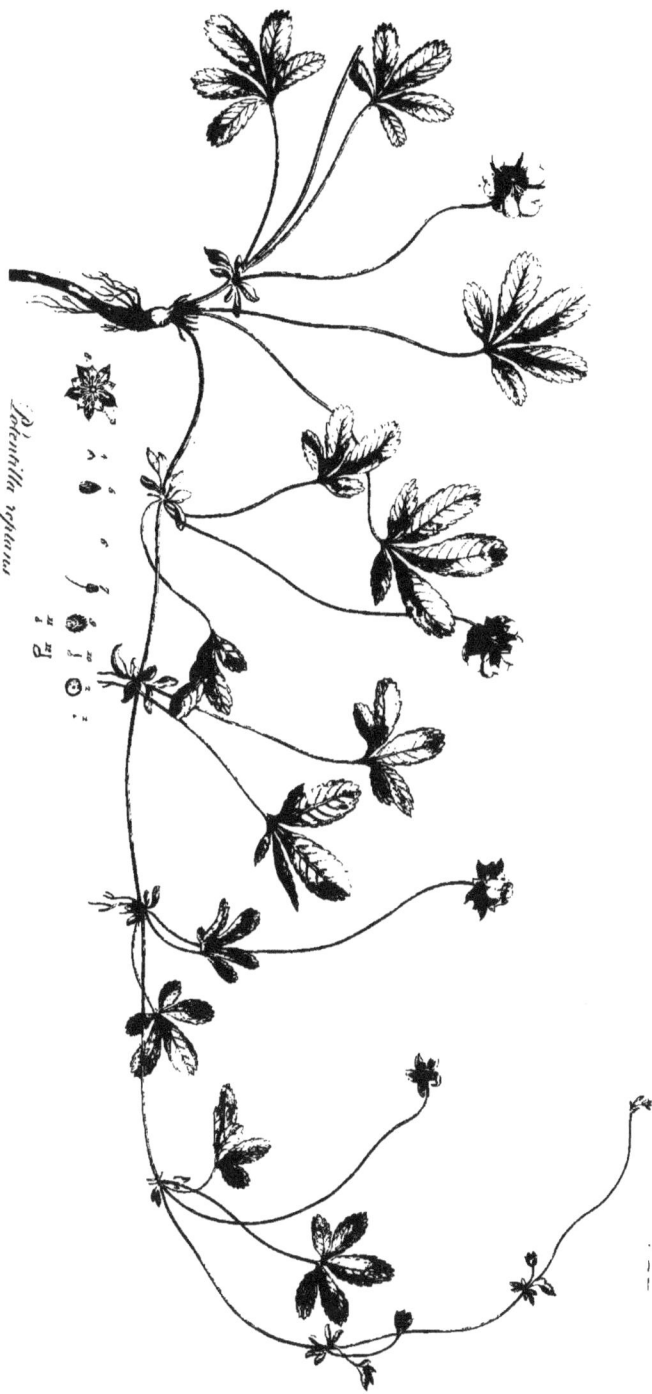

Potentilla reptans

Ranunculus bulbosus.

Ranunculus Bulbosus. Round-Rooted or Bulbous Crowfoot.

RANUNCULUS *Linnæi Gen. Pl.* Polyandria Polygynia.

Raii Syn: Gen. 15. Herbæ semine nudo polyspermæ.

RANUNCULUS *bulbosus*, calycibus retroflexis, pedunculis sulcatis, caule erecto multifloro, foliis compositis. *Linnæi Syst. Vegetab. p.* 430. *Sp. Pl.* 778. *Fl. Suecic.* 196.

RANUNCULUS radice subglobosa, foliis hirsutis, semitrilobis, lobis petiolatis acute serratis. *Haller. hist.* v. 2. p. 74.

RANUNCULUS *Scopoli Fl. Carn.* v. 1. p. 400. Diagn. Radix globosa. Calyces reflexi. Squamula nectarifera obtusæ trigona.

RANUNCULUS pratensis radice verticilli modo rotunda. *Bauhin. pin.* 179. *Fuschii Icon.* 160. *Gerard. emac.* 953. *Parkinson* 329. *Raii Synop.* 247. *Hudson Fl. Angl.* 211. *Fl. Dan. Icon.* 551.

RADIX perennis, *subrotunda*, albida, solida, superne et inferne depressior, hinc radicem Rapæ quodammodo referens.

ROOT perennial, *roundish*, white and solid, flattened a little both at top and bottom, hence somewhat resembling a Turnep.

CAULIS pedalis, teres, *erectus*, fistulosus, hirsutus, ramosus.

STALK a foot high, round, *upright*, hollow, hairy and branched.

FOLIA *radicalia* petiolis longis, hirsutis, basi vaginantibus insidentia, subprocumbentia, hirsuta, venosa, trilobata, lobo medio majori et longius petiolato, semitrifido, segmentis acute incisis; lobis lateralibus trifidis, segmentis inferioribus profundius divisis; caulina subsessilia in lacinias plures tenuiores divisa.

LEAVES: the *radical* leaves placed on long hairy footstalks, which at bottom embrace the stalk, somewhat procumbent, hairy, veiny and divided into three lobes; the mid-lobe largest and placed on a longer foot-stalk than the others, divided half way down into three segments which are sharply cut in; the side-lobes triid, the lower segments more deeply divided than the others; the leaves of *the stalk* nearly sessile, deeply divided into numerous and narrower segments.

PEDUNCULI *sulcati*.

FOOT-STALKS of the flowers *grooved*.

CALYX: Perianthium pentaphyllum, foliolis ovatis, concavis, *reflexis*, pilosis, apice obtusiusculis, margine membranaceis, basi *subpellucidis*, *fig.* 1.

CALYX: a Perianthium of five leaves, the leaves oval, hollow, *turned back* and hairy, bluntish at top, membranous at the edges, thin and somewhat *transparent* at bottom, *fig.* 1.

COROLLA Petala quinque obcordata, flava, nitentia, *fig.* 2.

COROLLA: five Petals, heart-shaped, yellow, and shining, *fig.* 2.

NECTARIUM: squamula flava submarginata ad basin petali *fig.* 3.

NECTARY: a small yellow scale at the bottom of the petal, with a slight indentation at top, *fig.* 3.

STAMINA: Filamenta plurima, receptaculo inserta; Antheræ oblongæ, flavæ, subincurvatæ, *fig* 4.

STAMINA: Filaments numerous and inserted into the receptacle; Antheræ oblong, yellow, and bending a little inwards, *fig.* 4.

PISTILLUM: Germina numerosa in capitulum collecta; Styli nulli; Stigmata minima reflexa, *fig.* 5.

PISTILLUM: Germina numerous, collected into a little head; Styles none; Stigmata very small and bending back, *fig.* 5.

SEMINA plurima compressa, fusca, mucronata, lævia, arillata, *fig.* 6.

SEEDS numerous, flat, brown, smooth, pointed, and covered with an Arillus, *fig.* 6.

Fig. 7, Arillus, *fig.* 8, semen denudatum.

Fig. 7, the Arillus, *fig.* 8. the seed taken out of it.

THIS *Crowfoot* has been considered by some Authors as the same Species with the *Ranunculus repens*, but certainly without any propriety, for there can be no doubt but they are as distinct as any two species of *Ranunculus* whatever. It is distinguished from the *repens* by several peculiarities, the principal of which are, 1st, its *reflexed calyx*, the turning back of which does not depend on any accidental circumstances, but solely on its particular structure; if it be plucked off, and held up to the light, the lower half of it will appear thin and almost transparent, hence not having a sufficient degree of solidity to support itself upright, it is reflected downwards;—2dly, the root in this species is *round*, and *solid*; in the *repens* it is *fibrous*; and 3dly, (which perhaps may be considered as the most essential difference) the stalk of the *bulbosus* is never known to *throw out any Stolones or Creepers*, which the *repens* always does in every soil and situation

This Species blows earlier than either the upright or creeping Crowfoot, and is the second flower, which next to the Dandelion covers our meadows and pastures with that delightful yellow, which almost dazzles the eye of the beholder.

Like the rest of the Crowfoots it possesses the property of inflaming and blistering the skin, but more particularly the Root, which is said to raise blisters with less pain and more safety, than Spanish flies; and hence where blisters have been thought necessary, these roots have been applied for that purpose, particularly to the Joints in cases of the Gout. On being kept they loose their stimulating quality, and are even eatable when boiled.

Hoffman informs us that Beggars make use of them to blister their skins in order to excite compassion.

The Juice of this herb is said to be more acrid than that of the *Ranunculus sceleratus*, and if applyed to the nostrils it provokes sneezing.

Hogs are fond of the roots and will frequently dig them up.

It abounds in dry pastures, and flowers in May; it is cultivated when double as well as the upright meadow Crowfoot, which last occurs in almost every Garden, under the name of *Yellow Batchelors Buttons*.

Ranunculus acris

Ranunculus acris. Upright Meadow Crowfoot.

RANUNCULUS *Linnæi Gen. Pl.* Polyandria Polygynia.

Raii Gen. 15. Herbæ semine nudo, polyspermæ.

RANUNCULUS *acris* calycibus patulis, pedunculis teretibus, foliis tripartito-multifidis, summis linearibus. *Linnæi Syst. Vegetab. p.* 430. *Fl. Suecic. p.* 196.

RANUNCULUS foliis hirsutis, semitrilobatis, lobis lateralibus bipartitis, foliis caulinis semitrilobis. *Haller. hist. n.* 1169.

RANUNCULUS pratensis erectus acris. *Bauhin. pin.* 178. *Gerard. emac.* 951. *Parkinson* 329. *Raii Synopsis, p.* 248. *Hudson. Fl. Angl. p.* 211. *Scopoli. Fl. Carniol. p.* 378.

RADIX perennis, e pluribus radiculis albidis constans.

ROOT perennial, consisting of numerous white fibres.

CAULIS bipedalis, erectus, fistulosus, teres, subpilosus, apice ramosus.

STALK generally about two feet high, upright, hollow, round, somewhat hairy, much branched at top.

FOLIA *Radicalia* petiolis longis erectis insidentia, tripartita, lobo medio trifido, lateralibus bilobis, omnibus acute dentatis aut incisis, subhirsutis, superne ad basin præsertim sæpe purpureis, venis subtus extantibus.
Caulina radicalibus similia, in lacinias tenuiores vero divisa et petiolis brevioribus insidentia, tandem linearia, sessilia. Petioli cum vaginis hirsuti.

LEAVES: *Radical* leaves standing on long upright foot-stalks, tripartite, the middle lobe triid, the side ones bilobous, and all of them sharply indented, slightly hirsute, the upper surface particularly at the base frequently of a purple colour, the veins underneath prominent.
Leaves of the *Stalk* like the radical leaves, but more finely divided, and standing on shorter foot-stalks, at top linear and sessile. The FOOTSTALKS with their sheaths hairy.

PEDUNCULI teretes.

FOOT-STALKS of the Flowers round.

CALYX: Perianthium pentaphyllum, patens, flavescens, pilosum, foliolis ovatis, concavis, obtusis, margine membranaceis, *fig.* 1.

CALYX: a Perianthium of five leaves, spreading, of a yellow colour and hairy, the leaves oval, concave, and membranous at the edges, *fig.* 1.

COROLLA: Petala quinque flava, nitentia, subcordata nunc emarginata, nunc integra, *fig.* 2.

COROLLA: five Petals, yellow and shining, nearly heart-shaped, sometimes notched, sometimes entire, *fig.* 2.

STAMINA: Filamenta plurima, apice paululum dilatata, *fig.* 5. 4. Antheræ flavæ, subincurvatæ, obtusæ, *fig.* 4.

STAMINA: Filaments numerous, a little dilated at top, *fig.* 5. 4. Antheræ yellow, obtuse, bending a little inward, *fig.* 4.

NECTARIUM: squamula emarginata, ad basin petalorum, *fig.* 3.

NECTARY: a small scale, slightly notched at top, at the base of each Petal, *fig.* 3.

PISTILLUM: Germina numerosa, in capitulum collecta, Styli nulli; Stigmata reflexa, *fig.* 6.

PISTILLUM: Germina numerous, forming a little head; Styles none, Stigmata reflex, *fig.* 6.

SEMINA: plurima, subrotunda, compressa, fusca, apice reflexa. *fig.* 7.

SEEDS numerous, roundish, flat, of a brown colour, bending back at the tip, *fig.* 7.

Most of the Ranunculi or Crowfoots are acrid and in some degree poisonous, and the species above described possesses this property in a very considerable degree; hence Linnæus has given it the name of *acris*; even pulling up the plant and carrying it to some little distance we have known sufficient to produce a considerable inflamation in the palm of the person's hand who held it. Cattle in general will not eat it, yet sometimes when they are turned hungry into a new field of Grafs, or have but a small spot to range in they will feed on it, and hence their mouths, as we have been credibly informed, have become sore and blistered. When made into hay it loses its acrid property, but is too stalky and hard to afford good Nourishment. It should seem therefore to be the interest of the Farmer as much as possible to root out this species from his Meadows that its place may be supplied with good sweet grass.

It grows too frequently in most of our meadows, and flowers in June and July.

The common people about Town and in many parts of the country call this and the other yellow Crowfoots by the names of *Butter-cups* and *Butter-flowers*, and this name seems to have originated from a supposition that the yellow colour of butter was owing to these plants; that this should be the case seems scarce probable, certainly it receives no good taste from it.

CALTHA PALUSTRIS. MARSH-MARIGOLD.

CALTHA *Linnæi Gen. Pl.* POLYANDRIA POLYGYNIA *Cal.* o. Petala quinque. Nectaria o. *Capsulæ* plures
polyspermæ.

Raii Syn. HERBÆ MULTISILIQUÆ SEU CORNICULATÆ.

CALTHA palustris. *Linnæi Syst. Vegetab. p.* 432. *Flor. Suecic.* 198.

CALTHA *Haller. hist. helv. p.* 32. *n.* 1188.

POPULAGO *palustris. Scopoli Fl. Carniol p.* 404.

CALTHA palustris flore simplici. *Bauhin pin* 276.

POPULAGO. *Tournefort. Tabernamont.*

CALTHA palustris vulgaris simplex. *Parkinson* 1213.

CALTHA palustris major. *Gerard. emac.* 817.

*Raii Syn.*272. Marsh Marigold. *Hudson Fl. Angl. p.* 214.

RADIX perennis, e plurimis fibris, teretibus, majusculis, albidis, constans.

ROOT perennial, consisting of numerous, round, large, white fibres.

CAULES ex eadem radice nascuntur plures, suberecti, pedales, fistulosi, pene teretes, glabri, ramosi, ad basin purpurei.

STALKS: several arise from the same root, almost upright, about a foot high, hollow, nearly round, smooth, branched, and purple at bottom.

FOLIA *radicalia* petiolata, cordato-reniformia, glabra, crenata, *caulina* subsessilia, ad apicem acutiora, et acute crenata.

LEAVES: the *radical* leaves placed on long foot stalks, betwixt an heart and kidney shape, smooth, shining, and notched or crenated; the leaves of the STALK nearly sessile, more pointed at top, and sharply crenated.

STIPULÆ fuscæ, membranaceæ, marcescentes.
RAMI dichotomi.
PEDUNCULI uniflori, erecti, fulcati.

STIPULÆ brown, membranous and withered.
BRANCHES dichotomous.
PEDUNCLES supporting one flower, upright, and grooved.

CALYX nullus.

CALYX wanting.

COROLLA: PETALA plerumque quinque, flava, magna, subrotundo-ovata, plana, patentia, superne non splendentia, *fig.* 1.

COROLLA generally consists of five large PETALS of a roundish oval shape and yellow colour, flat, spreading, and without any gloss on the upper side, *fig.* 1.

STAMINA: FILAMENTA numerosa, filiformia, Corollâ breviora, ANTHERÆ oblongæ, compressæ, incurvatæ, flavæ, *fig.* 2.

STAMINA: FILAMENTS numerous, filiform, shorter than the Corolla; ANTHERÆ oblong, flat, bending inward, and of a yellow colour, *fig.* 2.

PISTILLUM: GERMINA quinque ad decem, oblonga, compressa, erecta; STYLI nulli; STIGMATA simplicia, *fig.* 3.

PISTILLUM: GERMINA from five to ten, oblong, flattish, and upright; STYLES none; STIGMATA simple, *fig.* 3.

PERICARPIUM: CAPSULÆ totidem, acuminatæ, patentes, futurâ superiore dehiscentes, *fig.* 4.

SEED-VESSEL: so many CAPSULES as Germina, pointed, and spreading, opening at the superior future, *fig.* 4.

SEMINA plurima, subovata, pulchra, inferne olivacea, superne rufa, *fig.* 5.

SEEDS numerous, somewhat oval, beautifull, at bottom of an olive, and at top of a reddish colour.

LINNÆUS informs us that the *Caltha* is the first flower which proclaims the Spring in *Lapland*, and that it begins to blow about the end of May, with us it usually flowers in March and April, and last Spring, 1775, this plant was found in Blossom in the month of February, so remarkably forward was the Spring of that year.

It grows in wet Meadows and by the sides of Rivers, where it makes a very noble appearance, and when double, is often cultivated in Gardens, where it will grow very readily if the soil be favourable.

In the Country, Children collect it to ornament their Garlands on May day.

I scarce ever observed the leaves to be eaten by any animals, but the flowers are often destroyed by a species of CHRYSOMELA.

HALLER says that it is acrid and caustic and yet that it is eaten by Cows.

The flower Buds are pickled and used as Capers.

Caltha palustris.

Verbena Officinalis

VERBENA OFFICINALIS. VERVAIN.

VERBENA   *Lin. Gen. Pl.* DIDYNAMIA GYMNOSPERMIA.
            Raii Gen. 14. SUFFRUTICES, ET HERBÆ VERTICILLATÆ.
VERBENA   *officinalis*, tetrandra, spicis filiformibus, paniculatis: foliis multifido-laciniatis, caule solitario.
            Lin. Syst. Vegetab. p. 62.
VERBENA   foliis tripartitis rugosis, spicis nudis gracillimis *Haller. hist. v.* 1. *p.* 96.
VERBENA   communis cæruleo flore. *Bauhin, Pin.* 269. mas, seu recta et vulgaris. *Parkinson* 674. communis
            Gerard 664.   *Raii Syn.* 236.   *Hudson Fl. Angl. p.* 505.   *Scopoli Fl. Carniol. p.* 433.

RADIX perennis, lignosa, crassitie digiti minimi, raro major, in terram profunde penetrans, fibrosa, lutescens, sapore subamaro.

ROOT perennial, woody, about the thickness of the little finger, seldom larger, running deep into the earth, fibrous, of a yellowish colour, and slightly bitter taste.

CAULES plerumque plures ex eadem radice, erecti, pedales aut bipedales, quadrangulares, duo latera excavata, duo subconvexa, sulcata, idque alterna, aculeis brevibus armati, brachiati.

STALKS in general several arise from the same root, upright, from one to two feet high, four square, two sides hollowed out, two rotundish and grooved, and that alternately, armed with short prickles, the branches alternately opposite.

FOLIA opposita, sessilia, venosa, profunde dentata, aut incisa, ad basin angustiora.

LEAVES opposite, sessile, veiny, deeply indented or cut in, narrower at bottom.

FLORES in spicas longas, filiformes, erectas dispositi, BRACTEA ovato-lanceolata, acuminata, calyce breviore suffulti, *fig.* 11.

FLOWERS disposed in long filiform erect spikes, supported by an oval pointed FLORAL-LEAF shorter than the Calyx, *fig.* 11.

CALYX: PERIANTHIUM monophyllum, angulatum, quinquedentatum, *denticulo quinto minimo*, persistens, *fig.* 1, 2, 3.

CALYX: a PERIANTHIUM of one leaf, quinquedentate, *the fifth tooth exceedingly minute*, continuing, *fig.* 1, 2, 3.

COROLLA monopetala, inæqualis, purpurascens, TUBUS cylindraceus, incurvatus; FAUX villosa, *fig.* 5; LIMBUS quinquefidus, *laciniis* rotundatis, subæqualibus, *fig.* 4.

COROLLA monopetalous, unequal, purplish, the TUBE cylindrical and crooked, the MOUTH villous, *fig.* 5. the LIMB divided into five *segments*, which are round and nearly equal, *fig.* 4.

STAMINA: FILAMENTA quatuor brevissima, vix conspicua, ANTHERÆ quatuor, quarum duæ breviores reliquis, ejusdem formæ cum Didynamiis *fig.* 6.

STAMINA: four FILAMENTS very short and scarce conspicuous, four ANTHERÆ two of which are above the others, of the same form with those of the Class Didynamia in general, *fig.* 6.

PISTILLUM: GERMEN tetragonum, STYLUS filiformis apice paululum incrassatus; STIGMA obtusum *fig.* 7

PISTILLUM: the GERMEN four square, the STYLE filiform, growing thicker towards the extremity, the STIGMA obtuse, *fig.* 7.

PERICARPIUM nullum, Calyx continens Semina.

PERICARPIUM wanting, the Calyx containing the Seeds.

SEMINA quatuor, oblonga, obtusa, interne planiuscula alba, externe fusca, convexa, *sulcato-reticulata fig.* 8, 9, 10.

SEEDS four, oblong, obtuse, on the inside flatish and *white*, on the outside brown, convex, *grooved and reticulated, fig.* 8, 9, 10.

The Vervain may be considered as a kind of domestic plant, not confined to any particular soil, but growing by the road sides, pretty universally at the entrance into Towns and Villages.

It produceth its blossoms in the months of August and September.

There is only one Species of this Genus which grows wild in this country, but in different parts of the world the species are numerous, and what is remarkable, some have four and others but two Stamina, hence LINNÆUS ranks them among his *Diandrous* plants, making a division of them into such as have *flores Diandri* and *flores Tetrandri*. As our species hath four stamina, two of which are above the other two, as the Style proceeds from the center of the four united Germina, and as four naked seeds follow, which are contained within the Calyx, we have placed it with SCOPOLI among the *Didynamia Gymnospermia* plants, a Class to which the botanic Student, who had been instructed in the Linnæan principles of Botany, would readily have been induced to refer it.

The seed of this plant has something remarkably curious in its appearance, on the inside it is of a snowy white, externally brown, and beautifully reticulated.

The Plant which the Romans called *Verbena*, appears to have been used on particular occasions at a very early period, as a token of mutual confidence betwixt them and their Enemies. It was also constantly applied to the purposes of Superstition and Enchantment, in making wreaths and brooms for their Altars, and chaplets for their Priests. It is probable from *Pliny's* account, that the plant which we now describe was the same with that of the Antients, but in a larger sense, they called the Laurel and Myrtle or whatever was bound round the Altar *Verbena*. The dry harsh nature of this herb, agrees but ill with the *Pinguis Verbena* of Virgil, perhaps it acquired that title from being anointed with the fat of the sacrifice.

In later times Vervain has been accounted a sovereign remedy in a multitude of disorders; SCHRODER recommends it in upwards of thirty different complaints, on which Mr. Ray judiciously observes "*Mirum tot viribus pollere plantam nulla insigni qualitate sensibili dotatam*"! strange that a plant which inherits no remarkably sensible quality should possess so many virtues!

Mr. Morley a late writer on the Vervain, considers it as extremely useful in the cure of the Schrophula or Kings evil, and in his Essay on the nature and cure of Schrophulous diseases, has given us a figure of the plant with particular directions for its use, which consists in hanging the root (which is to be of a larger or smaller size according to the age of his Patients) tied with a yard of *white* sattin ribband round the neck, there to be worn till they recover.

Those who know any thing of the effects of Medicines on the human body, will not easily be persuaded that such a kind of application can produce any very wonderful effect in this case, even making the greatest allowance for the powers of the imagination; and Mr. Morley as if sensible of the inefficacy of his Vervain Amulet, calls to his assistance a number of powerful medicines, among others we find Mercury, Antimony, Hemlock, Jalap, &c; and by a repeated and oftimes a long continued application of Baths, Cataplasms, Ointments, Poultices, Plaisters, &c. and the exhibition of gentle purges and alterative medicines, some have been relieved and others cured; but can any one hence infer with any degree of reason that the Vervain Root had any share in the cure? certainly no; out of all Mr. Morley's cases there is not one which proves it, and the virtues of this plant still remain to be ascertained by rational experiments.

It should be observed that the Schrophula is a disease which at certain periods of life and at certain seasons of the year, is liable to be much worse than at others, and frequently exceeding bad cases of this kind have been cured by the most simple applications.

Many people have no doubt applied to Mr. Morley from a supposition that his motives were perfectly disinterested, and it must be confessed that there are Empirics much more mercenary and infinitely more dangerous; yet it does not appear but Mr. Morley acts nearly on the same principle with other Practioners in Physick, with this difference indeed, that they receive their fees in specie, he takes his in kind.

That we may not be thought to act disingenuously by Mr. Morley we shall quote his own words—"Many many Guineas have been offered me but I never take any money. Sometimes indeed genteel People have sent me small acknowledgements of Tea, Wine, Venison, &c. Generous ones, small pieces of Plate or other little Presents. Even neighbouring Farmers a Goose or Turkey, &c. by way of Thanks.

LAMIUM *Linnæi Gen. Pl.* Didynamia Gymnospermia. *Corollæ* labium fuperius integrum, fornicatum, labium inferius bilobum; faux utrinque margine dentata. *Lin. Defcrip. Gen. abbrev.*

Raii Syn. Gen. 14. Suffrutices et herbæ verticillatæ.

LAMIUM *purpureum* foliis cordatis obtufis petiolatis. *Linnæi Syft. Vegetab., p.* 446. *Sp. Pl.* 809. *Fl. Suecic.* 203.

LAMIUM foliis cordatis, obtufis, in fummo ramo congeftis. *Haller. inft. v.* 1. 118.

LAMIUM *purpureum. Scopoli Fl. Carniol. p.* 407. *n.* 701.

LAMIUM purpureum fœtidum, folio fubrotundo, five Galeopfis Diofcoridis. *Bauhin. pin.* 230. *Lamium rubrum. Gerard emac.* 703. *Parkinfon.* 604. *Raii. Synopfis* Small Dead Nettle or red Archangel 240. *Hudfon. Fl. Angl.* 225. *Oeder. Fl. Dan. icon.* 523.

RADIX annua, fibrofa.

CAULES plures, ad bafin debiles, et ramofi, prope fummitatem fere nudi, et fæpe colorati, femipedales, quadrangulares, fiftulofi, fcrabiufculi.

FOLIA oppofita, venofa; hirfutula, inferiora fubrotundo-cordata, crenata, longe petiolata : fuperiora ovato-cordata, obtufe ferrata, petiolis brevibus infidentia, alterne oppofita, reflexa, denfe et imbricatim congefta, et rubedine tincta.

FLORES purpurei, in fummis caulibus verticillatim denfius ftipati. Verticilli multiflori.

CALYX: Perianthium monophyllum, tubulatum, fuperne patentius, quinquedentatum, fubftriatum, hirfutulum, dentibus fubæqualibus, acuminatis. *fig.* 1.

COROLLA monopetala, ringens, pallide purpurea, *fig.* 2; tubus brevis, cylindraceus, *fig.* 6 ; faux inflata, margine utroque bidentata, *fig.* 4; denticulo fuperiori fpinæ fimili, inferiore obtufiore, maculâ notatâ ; labium fuperius, *fig.* 3, ovatum, concavum, villofulum, integrum, labium inferius bilobum, maculatum, lobis patentibus. *fig.* 5.

STAMINA: Filamenta quatuor, fubulata, alba, fub labio fuperiori recta, quorum duo longiora, *fig.* 7 ; Antheræ oblongæ, barbatæ, polline croceo repletæ. *fig.* 8.

PISTILLUM: Germen quadrifidum; Stylus filiformis, longitudine et fitu ftaminum; Stigma bifidum, acutum, *fig.* 9, 10, 11.

SEMINA 4 in fundo calycis, pallida, triangularia, apice truncata, marginata, *fig.* 12.

ROOT annual and fibrous.

STALKS feveral, at bottom weak and branched, near the top almoft naked, and frequently coloured, fix inches or more in height, quadrangular, hollow, and flightly rough.

LEAVES oppofite, veiny, flightly hairy, the lower ones of a roundifh-heart fhaped form, notched, and placed on footftalks, the uppermoft ones oval-heart-fhaped, obtufely ferrated, with fhort footftalks, alternately oppofite, growing thickly together, bent back and laying one over another, of a reddifh colour.

FLOWERS purple, growing thickly together on the tops of the ftalks in whirls; many flowers in each whirl.

CALYX : a Perianthium of one leaf, tubular, at top fpreading, with five teeth, fomewhat ftriated and hairy, the teeth nearly equal and long pointed. *fig.* 1.

COROLLA monopetalous, gaping, of a pale purple colour, *fig.* 2 ; the tube fhort and cylindrical, *fig.* 6 ; the entrance of the tube inflated, the margin on each fide furnifhed with two teeth, *fig.* 4; the uppermoft pointed like a thorn, the lowermoft blunter with a fpot on it; the upper lip *fig.* 3 ; oval, hollow, flightly villous, entire, the under lip divided into two lobes, fpreading a little from one another, and fpotted, *fig.* 5.

STAMINA: four Filaments, tapering and white, hid under the upper lip, two of which are longer than the reft *fig.* 7 ; the Antheræ oblong, bearded, and and full of a yellow pollen *fig.* 8.

PISTILLUM : Germen quadrifid; Stylus filiform, the length of the Stamina; Stigma bifid and pointed *fig.* 9, 10, 11.

SEEDS 4 in the bottom of the Calyx, of a pale brown, triangular, cut off as it were at top, with a margin round them, *fig.* 12.

Although this plant may perhaps with propriety be confidered as a Weed in Gardens, yet the bright colour of its tops and flowers, joined to its early appearance, contributes not a little to ornament our banks in the Spring, when few other plants appear in bloffom.

The Flowers are moft commonly of a bright red colour, fometimes white, and are much reforted to by Bees of various kinds.

The Leaves and Flowers are thofe parts of the plant, which are ufed in Medicine, although in the prefent practice they are fcarce regarded.

According to Linnæus it is boiled in *Upland*, a Province of *Sweden*, as a pot herb. A Variety of this plant occurs not unfrequently about Town, which has its leaves more deeply indented. Ray calls it *Lamium rubrum minus, foliis profunde incifis.* I have found it growing on a bank on the right hand fide of the way between *Pimlico* and *Chelfea.*

Lamium purpureum

Thymus acinos.

Thymus *Acinos.* Basil Thyme.

THYMUS *Linnæi Gen. Pl.* Didynamia Gymnospermia.

Calycis bilabiati faux villis clausa.

Raii Synop. Gen. 14. Suffrutices et Herbæ verticillatæ.

THYMUS *Acinos* caulibus adscendentibus, foliis dentato-serratis, calycibus basi ventricosis.

THYMUS *Acinos* floribus verticillatis, pedunculis unifloris caulibus erectis subramosis, foliis acutis, serratis. *Linn. Syst. Vegetab. p.* 452. *Flor. Suecic. p.* 209.

CLINOPODIUM foliis ovatis acutis serratis, flore foliis breviore. *Haller. hist. helv. n.* 237.

THYMUS *Acinos. Scopoli Fl. Carniol. p.* 426. *n.* 735.

CLINOPODIUM arvense ocimi facie. *Bauhin. pin. p.* 225.

CLINOPODIUM minus sive vulgare. *Parkinson.* 21.

OCYMUM sylvestre. *Gerard. emac.* 675.

ACINOS multis. *Bauhin. hist.* 32. 259. *Raii Syn. p.* 238. Wild Basil. *Hudson Fl. Angl. p.* 230.

RADIX annua, simplex, fibrosa.

CAULES adscendentes, semipedales, tetragoni, ramosi, hirsuti, purpurascentes; Rami cauli similes longi, patentes, imi oppositi.

FOLIA opposita, petiolata, ovato-acuta, medium interius petiolo proximum integrum, exterius mucroni proximum dentatum, margines paululum reflexi, ciliati, nervo medio venisque subtus hirsutis, superne vix hirsuta, impunctata, venis quam in serpyllo profundius exarata.

FLORES pedunculati, verticillati, spicati, plerumque sex in singulo verticillo.

CALYX: Perianthium monophyllum, tubulatum, basi ventricosum, striatum, hirsutum, quinquedentatum, dentibus tribus superioribus brevioribus, reflexis, inferioribus setaceis, fauce villis clauso, *fig.* 1.

COROLLA monopetala, tubulosa, purpurea, bilabiata, labium superius breve, obtusum, reflexum, emarginatum, inferius trifidum, laciniis subrotundis, medio productiore subemarginato, *macula alba, lunulata, prominente, notata, fig.* 3, 4, 5.

STAMINA: Filamenta quatuor, quorum duo longiora, Corollà breviora; Antheræ parvæ, rubræ, *fig.* 6.

PISTILLUM: Germen quadripartitum; Stylus filiformis longitudine Staminum; Stigma bifidum, acutum, *fig.* 7.

PERICARPIUM nullum

SEMINA quatuor oblonga intra Calycem, *fig.* 8, 9.

ROOT annual, simple and fibrous.

STALKS adscending, about six inches high, square, branched, hirsute, purplish; Branches like the stalk, long, spreading, the bottom ones opposite.

LEAVES opposite, standing on foot-stalks, of a pointed oval shape, the inner middle part of them next the foot-stalks entire, the outer middle part next the point indented, the edges turned a little back and ciliated, the midrib and veins on the under side of the leaf hirsute, the upper surface of the leaves scarcely hairy, without any dots, the veins deeper than in the common Wild Thyme.

FLOWERS growing on foot-stalks, in whirls, forming a spike, generally six in each whirl.

CALYX: a Perianthium of one leaf, tubular, bellying out at bottom, striated, hirsute, having five teeth, the three uppermost of which are shortest and turned back, the lower ones slender and tapering, the moth closed up with short hairs, *fig.* 1.

COROLLA monopetalous, tubular, purple, having two lips, the uppermost of which is shortest, blunt, turned back, with a slight notch in it; the lowermost divided into three roundish segments, the middle one of which is longer than the others, very slightly notched in, and marked with a *raised white semilunar spot, fig.* 3, 4, 5.

STAMINA: four Filaments, two long and two short, within the Corolla; Antheræ small and red, *fig.* 6.

PISTILLUM: Germen divided into four parts; Style filiform, the length of the Stamina; Stigma bifid and acute, *fig.* 7.

SEED-VESSEL none.

SEEDS. Four oblong seeds within the Calyx, *fig.* 8, 9.

As there are only two species of *Thyme* growing wild in this Kingdom, and those very different from each other, the young Botanist cannot be at a loss in distinguishing them; with the *Thymus alpinus*, (figured by that accurate Botanist Jacquin, in his *Fl. Austriac*, who has contributed much to the advancement of botanic knowledge,) this plant has a much greater affinity, but may be distinguished by attending to the size of the flowers and the shape of the Calyx: the flowers of the *alpinus* are nearly twice as large as those of the *acinos*, and the Calyx of the latter has a protuberance at its base which we do not find either in the *alpinus* or *serpyllum*; a white circular mark in the mouth of the flowers, makes the blossoms of this species strikingly different from those of Wild Thyme.

The most common place of growth for this plant is in uncultivated fields, particularly where the soil is chalky, about *Charlton* it is found in abundance, flowering in July and August.

A variety with a white flower sometimes occurs.

The same agreeable aromatic flavour predominates in this species as in the Wild Thyme, whence it is probable that their virtues are very similar.

Euphrasia Odontites.

EUPHRASIA *Linnæi Gen. Pl.* DIDYNAMIA ANGIOSPERMIA.

Raii Syn. Gen. HERBÆ FRUCTU SICCO SINGULARI FLORE MONOPETALO.

EUPHRASIA *Odontites* foliis linearibus: omnibus ferratis. *Linnæi Syst. Vegetab. Sp. Pl. p.* 841. *Fl. Suecic. p.* 213. *n.* 544.

ODONTITES bracteis ferratis hirfutis. *Haller. hist. v.* 1. *p.* 134. *n.* 304.

EUPHRASIA *Odontites. Scopoli Fl. Carniol. p.* 435.

EUPHRASIA pratenfis rubra. *Bauhin Pin. p.* 234.

EUPHRASIA pratenfis rubra major. *Parkinfon* 1329.

CRATÆOGONON Euphrofyne. *Ger. emac.* 91. *Raii Syn. p.** 284. Eye-bright Cow-wheat. *Hudfon Fl. Angl. p.* 234.

RADIX annua, fimplex, fibrofa, lignea.

ROOT annual, fimple, fibrous, and woody.

CAULIS erectus, ramofiffimus, femipedalis, ad bipedalem, hirfutus, obtufe quadrangularis.

STALK upright, very much branched, from fix inches to two feet high, hirfute, and obtufely fquare.

RAMI cauli fimiles, oppofiti.

BRANCHES like the ftalk and oppofite.

FOLIA alterne oppofita, feffilia, lineari-lanceolata, reflexa, rariter dentata, hirfutula, venofa, venis parvis, fubtus hirfutis.

LEAVES alternately oppofite, feffile, betwixt linear and lanceolate, turning back, thinly indented, flightly hirfute, veiny, veins few and hirfute underneath.

BRACTEÆ lanceolatæ, fuberectæ, purpurafcentes.

BRACTEÆ lanceolate, nearly upright, purplifh.

FLORES fpicati, fecundi, fpicis apice fubnutantibus,

FLOWERS growing in fpikes of a red colour, inclined all one way, the fpikes nodding a little at top.

CALYX : PERIANTHIUM monophyllum, tubulofum, quadridentatum, hirfutum, dentibus æqualibus, acutis, *fig.* 1.

CALYX : a PERIANTHIUM of one leaf, tubular, quadridentate, hirfute, the teeth equal and fharp, *fig.* 1.

COROLLA monopetala, ringens, labium fuperius concavum, fubemarginatum, inferius tripartitum, laciniis obtufis, æqualibus, *fig.* 2.

COROLLA monopetalous, gaping, the upper lip concave and flightly notched in ; the lower lip divided into three, obtufe, equal fegments, *fig.* 2.

STAMINA : FILAMENTA quatuor, quorum duo paulo breviora, alba; ANTHERÆ bilobæ, biloculares, apice filamentofæ, bafi fpinulis duabus terminatæ, deorfum ubi filamentum inferitur appendiculis clavatis pluribus inftructæ, *fig.* 3, 4, 5.

STAMINA : four FILAMENTS, two fomewhat longeft, white; ANTHERÆ compofed of two lobes and two cavities, at top thready, at bottom terminated by two little fpines, and on the back part where the filament is inferted, furnifhed with feveral fmall club-fhaped threads or appendages, *fig.* 3, 4, 5.

PISTILLUM : GERMEN ovatum, hirfutulum; STYLUS filiformis, in flore nondum explicato fub labio fuperiore Corollæ involutus, poftea Corollâ longior ; STIGMA capitatum, *fig.* 6.

PISTILLUM : GERMEN oval, hirfute ; STYLE filiform, before the flower opens bent in underneath the upper lip of the Corolla ; afterwards longer than the Corolla; STIGMATA forming a little head, *fig.* 6.

PERICARPIUM : CAPSULA ovato-oblonga, compreffa, bilocularis, *fig.* 7.

SEED-VESSEL an oval, oblong, flattifh CAPSULE, of two cavities, *fig.* 7.

SEMINA plurima, albida, ftriata, *fig.* 8.

SEEDS feveral, whitifh and ftriated, *fig.* 8.

This fpecies of *Eyebright*, which is exceedingly different from the common fort, grows very common in Paftures, fometimes in Corn-fields, and flowers in July and Auguft: it differs very much in fize according to the place it grows in, and is now and then found with white flowers.
It is not remarked either for its beauty or utility.

ANTIRRHINUM *CYMBALARIA.* IVY-LEAV'D ANTIRRHINUM.

ANTIRRHINUM *Linnæi Gen. Pl.* DIDYNAMIA ANGIOSPERMIA.

Raii Syn. HERBÆ FRUCTU SICCO SINGULARI FLORE MONOPETALO.

ANTIRRHINUM *Cymbalaria* foliis cordatis quinquelobis alternis, caulibus procumbentibus. *Linnæi Syft. Vegetab.* p. 464. *Sp. Pl.* p. 851.

ANTIRRHINUM caule repente, foliis reniformibus, quinquelobatis. *Haller hift.* p. 146. n. 339.

ANTIRRHINUM *Cymbalaria Scopoli Fl. Carniol.* n. 770.

CYMBALARIA *Bauhin pin.* 306.

LINARIA *hederaceo folio glabro, feu Cymbalaria vulgaris. Tourn.* 169. *Garidel.* 287. *Gouan. Fl. Monfp.* p. 100. *Gerard Fl. Galloprov.* p. 292. *Raii Syn.* p. *282. *Hudfon Fl. Angl.* p. 237.

Tota Planta glabra, cum odore ingrato.	The whole plant fmooth, with a difagreeable fmell.
RADIX perennis, fibrofa, intra fiffuras murorum penetrans; eradicatione difficilis.	ROOT perennial, fibrous, penetrating between the crevices of the walls, and fcarce to be eradicated.
CAULES plures, confertim nafcuntur, bafi repentes, procumbentes, ramofi, teretes, glabri, purpurafcentes, nervo intus duriore et tenaciore ficut in Alfine.	STALKS numerous, growing in a kind of tuft, creeping at bottom, procumbent, branched, round, fmooth, purplifh, and ftringy as in Chickweed.
FOLIA quinquelobata, glabra, fubcarnofa, oppofita, aut alterna, fæpe purpurafcentia, *fig.* 12.	LEAVES quinquelobate, fmooth, fomewhat flefhy, fome of them oppofite, others alternate, frequently purplifh, *fig.* 12.
PETIOLI longi, fuperne fulcati.	FOOT-STALKS of the leaves long, on the upper part grooved.
PEDUNCULI teretes, petiolis paulo longiores.	FOOT-STALKS of the flowers, round, a little longer than the foot-ftalks of the leaves.
CALYX: PERIANTHIUM quinquepartitum, laciniis lanceolatis, perfiftentibus, *fig.* 1.	CALYX: a PERIANTHIUM divided into five fegments, which are lanceolate and continuing, *fig.* 1.
COROLLA monopetala, ringens; TUBUS brevis, *fig.* 6; LIMBUS bilabiatus, labium fuperius bifidum, reflexum, purpureum, venis duabus faturatioribus ftriatum, *fig.* 2. inferius trifidum, laciniis fubrotundis, albidis, *fig.* 3; PALATUM prominens, bifidum, flavum, *fig.* 5, FAUX villofum, croceum.	COROLLA monopetalous, ringent; the TUBE fhort, *fig.* 6: the LIMB divided into two lips; the upper lip bifid, turning back, and purple, ftriped with two veins of a deeper colour, *fig.* 2; the lower lip trifid, the fegments round and whitifh, *fig.* 3; the PALATE prominent, bifid, and yellow, *fig.* 4; the MOUTH or entrance into the tube villous and faffron-coloured.
NEECTARIUM purpureum, conicum, longitudine calycis, *fig.* 5.	NECTARY purple, conical, the length of the Calyx, *fig.* 5.
STAMINA: FILAMENTA quatuor, duo breviora; ANTHERÆ bilobæ, albæ, conniventes, *fig.* 7.	STAMINA: four FILAMENTS, two fhort and two long; ANTHERÆ compofed of two lobes, white and connivent, *fig.* 7.
PISTILLUM: GERMEN fubrotundum, purpureum; STYLUS filiformis; STIGMA obtufum, *fig.* 8.	PISTILLUM: GERMEN roundifh and purple; STYLE filiform; STIGMA blunt, *fig.* 8.
PERICARPIUM CAPSULA fubrotunda, rugofa, feminibus protuberantibus, bivalvis, valvis apice in plures lacinias dehifcentibus, *fig.* 9, 10.	SEED-VESSEL a roundifh CAPSULE, furface uneven, from the feeds protuberating, of two valves, which open at top into feveral laciniæ, *fig.* 9, 10.
SEMINA nigra, fubrotunda, rugofa, *fig.* 10.	SEEDS black, roundifh and wrinkled, *fig.* 10.

This Species of *Antirrhinum* is fo perfectly diftinct from all the others which grow wild in this country, that there is no poffibility of miftaking it. It is found in great plenty in all thofe parts near London, that lay within the reach of the Thames; the feeds are carried by the flux and reflux of the tide up and down the river, and left at high water mark in the crevices of old walls, where they take root and encreafe very faft. It is fuppofed to have been introduced to us from Italy, whether for the purpofes of ornament or medicine is uncertain.

The Walls of the *Phyfic Garden*, at *Chelfea*, from whence it has probably originated in this country, are plentifully covered with it; it may alfo be found on the *Temple* Walls, and at the fides of the ftream running under *Vauxhall* Turnpike.

In fome fituations the leaves grow much larger than thofe of the annexed fpecimen.

Antirrhinum Cymbalaria.

ANTIRRHINUM *ELATINE*. SHARP-POINTED FLUELLIN.

ANTIRRHINUM *Lin. Gen. Pl.* DIDYNAMIA ANGIOSPERMIA.

Raii *Syn. Gen.* 18. HERBÆ FRUCTU SICCO SINGULARI, FLORE MONOPETALO IRREGU-
LARI.

ANTIRRHINUM foliis haſtatis alternis, caulibus procumbentibus. *Linn. Sp. Pl.* 85.

ANTIRRHINUM caule procumbente, foliis haſtatis, imis conjugatis, fuperioribus alternis. *Haller hiſt. v.* 1.
p. 14. 6. n. 340.

ELATINE folio acuminato, in baſi auriculato, flore luteo. *Bauhin Pin.* 253.

ELATINE folio acuminato. *Parkinſon* 553.

ELATINE altera. *Gerard emac.* 623.

LINARIA *Elatine* dicta, folio acuminato. *Raii Syn.* *282.

ANTIRRHINUM *Elatine*. Hudſon Fl. Angl. p. 237. Scopoli Fl. Carniol. p. 444. OEder. Fl. Dan. Ic. 426.

TOTA PLANTA piloſa.
RADIX fibroſa, annua, albida,
CAULES numeroſi, teretes, fubramoſi, in junioribus plantis fuberecti, tandem procumbentes, ad duos pedes et ultra ſæpe extenſi.

FOLIA petiolata, ima fubrotunda, oppoſita; proxima dentata, alterna; quæ ſequuntur magna ex parte haſtata.

PEDUNCULI axillares, alterni, penduli, longitudine foliorum.
CALYX: PERIANTHIUM quinquepartitum, perſiſtens, fegmentis ovato-lanceolatis acutis, fig. 1.
COROLLA monopetala, ringens, flava; TUBUS breviſſimus; LIMBUS bilabiatus, labium fuperius biſidum, fegmentis obtuſis, inferne purpureis, inferius trifidum, fegmentis obtuſis, medio productiore, et paulo minore; PALATUM prominulum, flavum, fig. 2; NECTARIUM fubulatum, flavum, longitudine fegmentorum calycis, fig. 3.

STAMINA: FILAMENTA quatuor, quorum duo paulo longiora; ANTHERÆ purpureo-fulcæ, coaleſcentes, fig. 4.
PISTILLUM: GERMEN fubrotundum, compreſſum, apice villoſum; STYLUS filiformis, longitudine ſtaminum, apice incraſſatus, uncinatus; STIGMA ſimplex, fig. 5, 6, 7.
PERICARPIUM: CAPSULA rotunda, bilocularis, bivalvis, valvis deciduis, foramine magno in utroque latere capſulæ relicto, valvæ orbiculatæ, concavæ, fig. 8, 9, 10.
SEMINA nigra, rugoſa, 8–10 in ſingulo loculamento, fig. 12.

THE WHOLE PLANT hairy.
ROOT fibrous, annual, whitiſh.
STALKS numerous, round, a little branched, in the young plants nearly upright, in the old ones trailing on the ground, frequently to the diſtance of two feet or more.

LEAVES ſtanding on foot-ſtalks, the bottom leaves roundiſh and oppoſite, the next to thoſe are indented and alternate, and thoſe which follow are for the moſt part haſtate.

PEDUNCLES alternate, pendulous, the length of, and proceeding from the Alæ of the leaves.
CALYX: a PERIANTHIUM divided into five ſegments perſiſting, the ſegments lanceolate, fig. 1.
COROLLA monopetalous, ringent, and yellow; the TUBE very ſhort; the LIMB divided into two lips, the upper lip bifid, the ſegments obtuſe, and purple underneath; the lower lip trifid, the ſegments obtuſe, the middle one longeſt and leaſt; the PALATE prominent and yellow, fig. 2; the NECTARIUM the length of the ſegments of the Calyx, ſmall and tapering, fig. 3.

STAMINA four FILAMENTS, two of which are a little longer than the others; the ANTHERÆ purpliſh-brown, adhering together, fig. 4.
PISTILLUM: the GERMEN roundiſh, flattened, at top hairy; the STYLE filiform, the length of the Stamina, thickened at top and hooked; the STIGMA ſimple, fig. 5, 6, 7.
SEED-VESSEL: a round CAPSULE of two cavities and two valves, the valves round and concave, on falling off leaving a large hole in each ſide of the Capſule, fig. 8, 9, 10.
SEEDS black, and wrinkled, from 8 to 10 in each cavity, fig. 12.

THIS ſpecies of *Antirrhinum* grows generally in Corn-fields, and in ſome parts of England is much more common than it is with us; in the Corn-fields about Peckham I have generally found it in bloom in July, Auguſt and September, and even later; it very much reſembles the *Antirrhinum ſpurium* in its general habit, but is readily diſtinguiſhed by its *pointed leaves*. Some Writers have conſidered it as poſſeſſed of healing properties, and affirm that the expreſſed juice of the plant, or its diſtilled water taken inwardly and applied externally, has checked and cured ſpreading and cancerous Ulcers; and RAY relates a Story from LOBEL of a poor Barber, who by the above uſe of this plant, ſaved his Noſe, which had been condemned to be cut off by ſeveral eminent Phyſicians and Surgeons.

Antirrhinum Elatine.

Antirrhinum *Linaria.* Common yellow

Toad Flax.

ANTIRRHINUM *Linnæi Gen. Pl.* DIDYNAMIA ANGIOSPERMIA.

Raii Syn. Gen. 18. HERBÆ FRUCTU SICCO SINGULARI FLORE MONOPETALO.

ANTIRRHINUM *Linaria* foliis lanceolato-linearibus confertis, caule erecto, fpicis terminalibus feffilibus, floribus imbricatis. *Linn. Syft. Vegetab. p.* 466. *Fl. Suecic.* 217.

ANTIRRHINUM foliis linearibus adfcendentibus congeftis, caule erecto fpicato. *Haller. hift. V.* 1. *p.* 145.

LINARIA vulgaris lutea flore majore. *Bauhin pin. p.* 212.

LINARIA lutea vulgaris. *Gerard emac.* 550. vulgaris noftras. *Parkinfon* 458. *Raii Syn. p.* *281. *Hudfon*

Fl. Angl. p. 238. *Scopoli Fl. Carniol. p:* 442:

RADIX perennis, alba, dura, lignofa, per terram reptando immenfûm fe propagans.

ROOT perennial, white, hard and woody, creeping under the earth, and propagating itfelf very much.

CAULES plerumque plures ex eadem radice, erecti, pedales aut cubitales, foliofiffimi, teretes, læves.

STALKS: generally feveral arife from the fame root, upright, from one to two feet high, very full of leaves, round and fmooth.

FOLIA linearia, acuta, conferta, fparfa, glauca.

LEAVES linear, pointed, growing very thick together on the ftalk without any regular order, fmooth, and of a blueifh colour.

FLORES lutei, palato croceo, in fummis caulibus in fpicas denfas imbricatim congefti.

FLOWERS yellow, with the palate of an orange or faffron colour, placed one over another in thick fpikes on the top of the Stalks.

CALYX: PERIANTHIUM quinquepartitum, breve, perfiftens, laciniis ovato-lanceolatis, fuperiore cæteris paulo longiore, duabus inferioribus magis dehifcentibus, *fig.* 1.

CALYX: a PERIANTHIUM divided into five fegments fhort and continuing, the fegments oval and pointed, the upper one a little longer than the reft, the two inferior ones gaping wideft, *fig.* 1.

COROLLA monopetala ringens, lutea, *fig.* 3. TUBUS brevis; LIMBUS bilabiatus, *fig.* 4. LABIUM *fuperius* bifidum, laciniis primum deflexis, poftea reflexis conniventibus, *fig.* 5; LABIUM *inferius* trifidum, laciniis obtufis, intermedio breviore minore, *fig.* 6; FAUX claufa PALATO prominente, bifido, croceo, ad bafin villofo, *fig.* 7.

COROLLA monopetalous, ringent, and yellow, *fig.* 3; the TUBE fhort; the LIMB compofed of two LIPS, *fig.* 4; the *upper* LIP bifid, the fegments firft bending down, afterwards turned back and clofing together, *fig.* 5; the *lower* LIP trifid, the fegments obtufe, the middle one fhorteft and leaft, *fig.* 6; the MOUTH clofed by a PALATE prominent, bifid, of a faffron colour, and villous at bottom, *fig.* 7.

STAMINA: FILAMENTA quatuor, alba, fub labio fuperiori inclufa, quorum duo breviora, ad bafin villofa, *fig.* 9; ANTHERÆ flavæ, conniventes, *fig.* 10.

STAMINA: four white FILAMENTS, inclofed under the upper lip of the Corolla, two of which are fhorter than the other two, at bottom villous, *fig.* 9; ANTHERÆ yellow, flightly connected together, *fig.* 10.

PISTILLUM: GERMEN fubrotundum, STYLUS filiformis, albus; STIGMA obtufum.

PISTILLUM· GERMEN roundifh, STYLE filiform and white; STIGMA obtufe.

PERICARPIUM: CAPSULA ovato-cylindracea, bilocularis, apice in plures lacinias dehifcens, *fig.* 14, 15, 16.

SEED-VESSEL a CAPSULE of an oval and cylindrical fhape, having two cavities, and fplitting at top into feveral divifions, *fig.* 14, 15, 16.

SEMINA numerofa, nigra, plana, medio extuberantia, *fig.* 17.

SEEDS numerous, black, flat, protuberant in the middle, *fig.* 17.

Mr. Ray in his *Hiftoria Plantarum* has collected the Authorities of feveral writers who fpeak highly of the medical virtues of this Plant. At the fame time that we by no means believe in all the Virtues which are attributed to many plants by the old Authors, we would be carefull of rejecting all their accounts, particularly when there is fome reafon to think they may be founded in Truth, the mention of them may at leaft ferve to excite fuch of the Faculty as have proper opportunities to give them a fair trial, and either reject them entirely, or bring them more generally into practice.

According to fome it operates both by Stool and Urine, and fo much by the latter, as to acquire among the *Germans* the name of *Harnkrout.* A fmall Glafs of the diftilled Water mixed with a drachm of the bark of the *Ebulus* or *Water Elder* in powder. powerfully provokes Urine, and is recommended in Dropfical Cafes. The diftilled water or juice of the Plant put in the Eyes, takes away the rednefs and inflammation of them, as TRAGUS afferts from his own long obfervation and experience. Made into an Ointment with lard and mixed with the yolk of Egg, it takes away the violent pain arifing from the Piles.

The flowers of this plant are frequently found double with two or more Spurs, and a fingular variety of it which LINNÆUS calls *Peloria*, is faid by Mr. HUDSON to grow about *Clapham* in Surry, this rare monftrofity we fhall not fail to figure.

In its common ftate, the Toad Flax grows very common on banks by the road fides, which it decorates not a little by its fingular and beautiful Flowers. It may with the greateft cafe be cultivated in Gardens, and raifed either from Seeds or Roots; the Seed is ripe at the latter end of September.

Antirrhinum Linaria.

DIGITALIS PURPUREA. FOX-GLOVE.

DIGITALIS *Linnæi Gen. Pl.* DIDYNAMIA ANGIOSPERMIA.

Cal. 5-partitus. *Cor.* campanulata 5-fida, ventricofa. *Caps.* ovata bilocularis.

Raii Syn. Gen. 18. HERBÆ FRUCTU SICCO SINGULARI FLORE MONOPETALO.

DIGITALIS *purpurea* calycinis foliolis ovatis acutis, corollis obtufis : labio fuperiore integro. *Linn. Syft.*

Vegetab. p. 570. *Sp. Pl. p.* 866.

DIGITALIS foliis calycinis ovatis, galea fimplici. *Haller. hift. p.* 143. *n.* 330.

DIGITALIS *purpurea. Scopoli Fl. Carniol. p.* 447. *n.* 780.

DIGITALIS *purpurea* folio afpero. *p.* 243.

DIGITALIS *purpurea. Gerard. emac.* 790.

DIGITALIS *purpurea* vulgaris. *Parkinfon* 1653. *Raii. Syn. p.* 283. Purple Fox-glove. *Hudfon Fl.*

Angl. p. 240. *Order Fl. Dan Icon.* 774.

RADIX biennis, fibrofa.

CAULIS tripedalis ad orgyalem, fimplex, erectus, foliofus, teres, pubefcens.

FOLIA ovato-acuta, ferrata, venofa, fubtus albida, pubefcentia ; PETIOLI breves, alati.

FLORES fpicati, nutantes, imbricati, fecundi.

PEDUNCULI uniflori, pubefcentes, apice incraffati, peracta florefcentiâ fuberecti.

CALYX: PERIANTHIUM quinquepartitum, laciniis ovato-acuminatis, nervofis, fupremâ anguftiore, *fig.* 1.

COROLLA monopetala, fubcampanulata, purpurea, interne ocellata ; TUBUS magnus, patens, deorfum ventricofus, bafi cylindracea, arcta; LIMBUS parvus, quadrifidus, laciniâ fuperiore integra, quafi truncata, inferiore majore, inflexa.

STAMINA: FILAMENTA quatuor bafi Corollæ inferta, alba, apice paululum latiora, bafi infracta, quorum duo longiora ; ANTHERÆ primum magnæ, turgidæ, ovatæ, bafi coadunatæ, lutefcentes, et fæpe maculatæ ; demum et formâ et fitû mire mutantur, *fig.* 2, 3, 4.

PISTILLUM : GERMEN fubconicum, luteo-virens ; STYLUS fimplex ; STIGMA bifidum, *fig.* 5, 6, 7.

NECTARIUM GLANDULA bafin Germinis cingens, *fig.* 8.

PERICARPIUM : CAPSULA ovato-acuminata, bilocularis, bivalvis, valvulâ inferiore findente, *fig.* 9.

SEMINA plurime, nigricantia, parva, utraque extremitate truncata, *fig.* 10.

ROOT biennial and fibrous.

STALK from three to fix feet high, fimple, upright, leafy, round, and pubefcent or downy.

LEAVES of a painted oval fhape, ferrated, veiny, underneath whitifh and pubefcent ; the FOOT-STALKS fhort and winged.

FLOWERS growing in a fpike, pendulous, laying one over another all one way.

PEDUNCLES fuftaining one flower, pubefcent, thickeft at top, after the flower drops off, becoming nearly upright.

CALYX: a PERIANTHIUM divided into five fegments, which are of an oval pointed fhape, and nervous, the uppermoft narrower than the reft, *fig.* 1.

COROLLA monopetalous, fomewhat bell-fhaped, purple, and marked in the infide with little eyes; the TUBE large, fpreading, bulging out backwards; the bafe cylindrical, and as if it had been tyed with a ligature ; the LIMB fmall and quadrifid, the upper fegment entire and as if cut off, the lower fegment larger and bent inward.

STAMINA: four FILAMENTS inferted into the bottom of the Corolla, white, a little broadeft at top, crooked at bottom, two long and two fhort ; ANTHERÆ at firft large, turgid, oval, touching at bottom, of a yellowifh colour and often fpotted ; laftly changing both their form and fituation in a fingular manner, *fig.* 2, 3, 4.

PISTILLUM : GERMEN rather conical, of a yellow green colour; STYLE fimple; STIGMA bifid, *fic.* 5, 6, 7.

NECTARY a GLAND furrounding the bottom of the Germen, *fig.* 8.

SEED-VESSEL: a pointed oval CAPSULE, of two cavities and two valves, the lowermoft valve fplitting in two, *fig.* 9.

SEEDS numerous, blackifh, fmall, as if cut off at each end, *fig.* 10.

Was it not that we are too apt to treat with neglect the beautiful plants of our own country, merely becaufe they are common and eafily obtained, the ftately and elegant *Fox-glove* would much oftener be the pride of our gardens than it is at prefent ; for it is not only peculiarly ftriking at a diftance, but its flowers and their feveral parts become beautiful in proportion to the nearnefs of our view : How fingularly and how regularly do the bloffoms hang one over another ! How delicate are the little fpots which ornament the infide of the flower ! and like the wings of fome of our fmall Butterflies fmile at every attempt of the Painter to do them juftice : how pleafing is it to behold the neftling Bee hide itfelf in its pendulous bloffoms ! while extracting its fweets which furnifh our tables with honey, and our manufacturers with wex : nor are the more interior parts of the flower lefs worthy of our admiration, or lefs adapted to the improvement of the young Botanift : here all the parts of the fructification being large, he will readily obtain a diftinct idea of them ; but more particularly of the form of the Antheræ, and the alteration which takes place in them, previous to and after the difcharge of the Pollen. *vid. fig.* 3, 4.

The flowers of this plant are in general of a fine purple colour, and like all other purple flowers are liable to variations ; fometimes we find the bloffoms of a milk white or cream colour, and fome other varieties of it are mentioned by RAY, but the white is the moft common. Such as would wifh to cultivate it, may raife it either from feed, which is very fmall for the fize of the plant, or from young plants. It grows naturally in a dry and gravelly foil, and in fuch fituations is common enough over moft parts of England ; about *Charlton-Wood* it is very plentiful, and flowers in July and Auguft.

According to the teftimony of many writers, the juice or decoction of this plant taken inwardly, acts as an emetic and purgative, and that too with confiderable violence; hence Mr. RAY very properly advifes it to be given to fuch only as have robuft conftitutions. PARKINSON affirms that it is very efficacious in the cure of the Epilepfy ; but he unites with it in his prefcription *Polypody of the Oak*, fo that there is no knowing to which of the plants the merit of curing this ftubborn difeafe is due.

The flowers or herb either bruifed or made into an ointment, are ftrongly recommended in Schrophulous tumours and ulcers ; and fo great an opinion have the Italians of its virtues as a vulnerary, that they have the following proverb concerning it. " *Aralda tutte le piaghe falda.*" Fox-glove cures all wounds. *Raii Hift. Plant.*

Digitalis purpurea

Draba verna. Vernal Draba or Whitlow Grass.

DRABA *Linnæi Gen. Pl.* TETRADYNAMIA SILICULOSA.

Raii Synop. Gen. 21. HERBÆ TETRAPETALÆ SILIQUOSÆ ET SILICULOSÆ.

DRABA *verna* fcapis nudis, foliis fubferratis. *Linnæi Syft. Vegetab. p.* 489. *Flor. Suec. p.* 223.

DRABA cauliculis nudis, foliis fubhirfutis, fubdentatis. *Haller. hift. helv.* 1. 215.

BURSA PASTORIS minor loculo oblongo. *Bauhin. pin.* 108. 2.

PARONYCHIA vulgaris. *Gerard emac.* 624. *Raii Syn.* 292: *Hudfon Fl. Angl.* 243. *Scopoli Flor. Carniol. n.* 792.

RADIX fibrofa, annua.

ROOT fibrous and annual.

CAULES nudi, palmares, 1 ad 5 aut plures in folo fertili ex eadem radice nafcuntur.

STALKS naked, about three inches high, one to five and frequently more, if the foil be rich, fpring from the fame root.

FOLIA ovato-lanceolata, bafi anguftiora integra et fubferrata, (ferra nifi unica aut duo, raro plures) fuper terram expanfa, fcabriufcula, hirfuta, pili bi-trifurci.

LEAVES of an oval pointed fhape, narrower at bottom, fome of them entire, and others a little ferrated, or indented, (feldom more than one or two indentations in a leaf,) fpreading on the ground, roughifh, hirfute, fome of the hairs bifurcate, others trifurcate.

PEDUNCULI alterni, uniflori.

PEDUNCLES alternate, uniflorous.

CALYX: PERIANTHIUM tetraphyllum, foliolis erectis, concavis, gibbis, obtufis, fubhirfutis. *fig.* 1.

CALYX: a PERIANTHIUM of four leaves, which are upright, hollow, gibbous, obtufe, and fomewhat hairy, *fig.* 1.

COROLLA tetrapetala, petala alba, calyce duplo longiora, bipartita. *fig.* 2.

COROLLA tetrapetalous, the PETALS white, twice the length of the Calyx, and bipartite, *fig.* 2.

STAMINA: FILAMENTA fex incurvata, quorum 4 longitudine Piftilli 2 breviora; ANTHERÆ flavæ. *fig.* 3. 4.

STAMINA: fix FILAMENTS which bend inward, 4 long the height of the Piftillum, and 2 fhort; the ANTHERÆ yellow, *fig.* 3. 4.

PISTILLUM: GERMEN ovatum, compreffum; STYLUS vix ullus; STIGMA capitatum, planum. *fig.* 5.

PISTILLUM: The GERMEN oval and flat; STYLE fearce any; STIGMA a fmall head flat at top.

PERICARPIUM: SILICULA ovata, compreffa, brevi mucrone obtufo terminata, bilocularis, bivalvis, valvulis plano-concavia. *fig.* 6.

SEED-VESSEL a fhort oval pod, flat, and terminated by a fhort blunt point, having two Cavities and two Valves, the Valves flightly concave, *fig.* 6:

SEMINA plura, ovata, fufca, margini DISSEPIMENTI affixa. *fig.* 8. 9.

SEEDS feveral, oval, brown, fixed to the edge of the DISSEPIMENT or Partition, *fig.* 8. 9.

ON Walls, dry Banks, and in barren Fields, the white bloffoms of this diminutive plant, are very confpicuous in the months of March and April, a feafon when any kind of bloffom is viewed with pleafure, as it cannot fail to excite the pleafing reflection that the feafon is approaching when

> " *All that is fweet to fmell, all that can charm*
> *Or eye or ear, burfts forth on every fide*
> *And crouds upon the fenfes.*"

Linnæus informs us that in Smoland a Province of Sweden, they fow their Rye when this plant is in bloffom, and that in the night time and in wet weather its flowers droop.

Galen fays that *Paronychia* or *Whitlow Grafs* has its name from its properties, for it heals Whitlows; but Commentators are much in doubt concerning the plant itfelf. From the account of the Antients, it appears that it is a different plant from what we are now defcribing; fome have fixed on Wall Rue, (ASPLENIUM *Ruta muraria*,) others on a plant refembling Spurge, fuch is the confufion that arifes from imperfect defcriptions.

Draba verna

Thlaspi Bursa Pastoris.

THLASPI BURSA PASTORIS.   SHEPHERD's PURSE.

THLASPI *Linnæi Gen. Pl.* TETRADYNAMIA SILICULOSA.

Silicula emarginata, obcordata, polyſperma : valvulis navicularibus, margi-

nato-carinatis.

Raii Syn. Gen 21. HERBÆ TETRAPETALÆ SILIQUOSÆ ET SILICULOSÆ.

THLASPI *Burſa paſtoris* ſiliculis obcordatis, foliis radicalibus pinnatifidis. *Linnæi Syſt. Vegetab. p.* 491.

Spec. Pl. 903. *Fl. Suecie.* 227.

NASTURTIUM ſiliquis triangularibus, *Haller hiſt. v.* 1. *p.* 221

PASTORIA BURSA *Fuſchii icon.* 611.

BURSA PASTORIS major folio ſinuato. *Bauhin Pin.* 108. *Gerard emac.* 276. *Parkinſoni Theat.* 866.

Raii Syn. 306. *Hudſon, Fl. Angl.* 247. *Scopoli. Fl. Carniol. v.* 2. 17.

RADIX annua, fibroſa, albida.

ROOT annual, fibrous and whitiſh.

CAULIS pedalis, erectus, ramoſus, teres, ſubaſper.

STALK about a foot high, upright, branched, round, a little prickly.

FOLIA *radicalia* hirſutula, pinnatifida, laciniis quoad formam mire variantibus, *caulina* amplexicaulia, dentata.

LEAVES : *radical leaves* ſlightly hirſute, pinnatifid, the laciniæ or jags varying exceedingly in their form ; the upper leaves embracing the ſtalk, and indented at the edges.

PEDUNCULI uniflori, demum ſere horizontales.

PEDUNCLES, ſupporting one flower on each, nearly horizontal when the flowers are gone off.

CALYX : PERIANTHIUM tetraphyllum, foliolis ovatis, concavis, ſubpiloſis, margine membranaceis, *fig.* 1.

CALYX : a PERIANTHIUM of four leaves, the leaves oval, hollow, ſlightly hairy, and membranous at the edges, *fig.* 1.

COROLLA : PETALA quatuor alba, calyce paulo longiora, apice rotundata, *fig.* 2.

COROLLA : four white PETALS, a little longer than the Calyx, round at top, *fig.* 2.

STAMINA : FILAMENTA ſex, alba, quorum quatuor longitudine Styli, duo breviora incurvata ; ANTHERÆ flavæ, *fig.* 3.

STAMINA : ſix white FILAMENTS, four of which are of the ſame length as the Style ; two are ſhorter and bent a little inwards : ANTHERÆ yellow, *fig.* 3.

PISTILLUM : GERMEN oblongo-cordatum ; STYLUS breviſſimus ; STIGMA villoſum, *fig.* 4.

PISTILLUM :GERMEN of an oblong heart-ſhape ;STYLE very ſhort ; STIGMA villous, *fig.* 4.

PERICARPIUM : SILICULA lævis, *obcordata*, bivalvis, *fig.* 5.

SEED-VESSEL ; a ſhort ſmooth pod, triangular or *heart-ſhaped*, with two valves, *fig.* 6.

SEMINA plurima, pedicellata, flaveſcentia, margini Diſſepimenti affixa, *fig.* 6.

SEEDS numerous, of a yellowiſh colour, ſtanding on little foot-ſtalks, which connects them to the edge of the Diſſepimentum or Partition, *fig.* 6.

DISSEPIMENTUM utrinque acutum Valvis contrarium.

PARTITION pointed at both ends, placed croſs-ways to the Valves.

THE radical leaves of this plant differ ſo exceedingly in their appearance, that the moſt expert Botaniſt is often obliged to have recourſe to its moſt ſtriking character, the ſhape of its Seed-veſſels, before he can with certainty diſtinguiſh it. When it grows on walls and in dry ſituations, the leaves are more deeply divided, and the Laciniæ become much narrower; in cultivated ground they are broader and leſs jagged : It differs likewiſe no leſs with reſpect to its ſize, ſometimes being not more than two or three inches high, and at other times as many feet.

March and *April* are the months in which it is found moſt generally in bloſſom, yet like the *Groundſel* and *Pea annua*, it may be found in this ſtate at almoſt any time of the year,

It acquires its name of *Shepherd's Pouch* or *Purſe*, from the particular ſhape of its pods, by which it is obviouſly diſtinguiſhed from all our other Tetradynamous plants.

The plant is collected and given to ſmall birds, who appear to be very fond of the ſeeds, and this is the only uſe to which we at preſent know of its being applied.

GERANIUM CICUTARIUM. HEMLOCK-LEAV'D CRANE'S-BILL.

GERANIUM *Linnæi Gen. Pl.* MONADELPHIA DECANDRIA.

Monogyna. *Stigmata* quinque. *Fruſtus* roſtratus, pentacoccus.

Raii Synop. HERBÆ PENTAPETALÆ VASCULIFERÆ.

GERANIUM *cicutarium* pedunculis multifloris, floribus pentandris, foliis pinnatis inciſis obtuſis, caule ramoſo. *Linnæi Syſt. Vegetab.* p. 90. *Fl. Suecic.* p. 243.

GERANIUM petiolis multifloris, caule procumbente, foliis duplicato-pinnatis, pinnulis acute inciſis. *Haller hiſt. No.* 944.

GERANIUM cicutæ folio minus, et ſupinum. *Bauhin pin.* 319.

GERANIUM cicutæ folio inodorum album. *Gerard emac.* 945. 946.

GERANIUM moſchatum inodorum. *Parkinſon* 1708. *Raii Syn.* 357. Field Crane's-bill without ſcent. *Hudſon Fl. Angl.* 262.

RADIX annua, albida, ſimplex, carne tenera, cum nervo intus duriore et tenaciore, paucis fibris inſtructa, craſſiuſcula, et in terram alte deſcendens.

ROOT annual, whitiſh, ſimple, tender, the ſtring or nerve in the middle of it hard and tough, furniſhed with few fibres, large for the ſize of the plant, and penetrating deep into the earth.

CAULES ex eadem radice naſcuntur plures, craſſiuſculi, teretes, hirſuti, procumbentes, ramoſi, variæ longitudinis pro ratione loci.

STALKS ſeveral uſually ſpring from the ſame root, thickiſh, round, hirſute, procumbent and branched, of various lengths according to their place of growth.

FOLIA pinnata, pinnis ſeſſilibus pubeſcentibus, pinnulis acute inciſis.

LEAVES pinnated, the pinnæ ſeſſile and ſlightly hairy, the pinnulæ ſharply indented.

STIPULÆ ad exortum foliorum membranaceæ, albidæ, ovato-acutæ, ſuperiore integra, *fig.* 1; inferiore in duas diviſæ, *fig.* 2.

STIPULÆ at the baſe of the leaves membranous, whitiſh, acutely oval, the uppermoſt intire, *fig.* 1; the lowermoſt generally divided into two, *fig.* 2.

PEDUNCULI axillares, alterni, hirſuti, multiflori, longitudine foliorum.

FOOT-STALKS of the flowers ſpringing from the baſe of the leaves, alternate, hirſute, the length of the leaves, and ſupporting many flowers.

FLORES umbellati, roſei, a tribus ad ſex.

FLOWERS growing in an Umbell, from three to ſix, of a roſe-colour.

INVOLUCRUM membranaceum, multidentatum, *fig.* 3; PEDICELLI baſi craſſiores, deflexi et demum aſſurgentes.

INVOLUCRUM membranous, with many teeth, *fig.* 3: the ſmall foot-ſtalks of the flowers thickeſt at bottom, turning down, and laſtly turning upward.

CALYX: PERIANTHIUM pentaphyllum, foliolis ovatis, ſtriatis, hirſutis, concavis, mucronatis, *fig.* 4.

CALYX: a PERIANTHIUM of five leaves, the folioli oval, ſtriated, hirſute, concave, and terminating in a fine point, *fig.* 4.

COROLLA: PETALA quinque, ſubovata, plana, ſubæqualia, roſea, baſi hirſuta, calyce longiora, *fig.* 5.

COROLLA: five PETALS, ſomewhat oval, flat, nearly equal, of a roſe colour, hairy at bottom, ſomewhat longer than the Calyx, *fig.* 5.

STAMINA: FILAMENTA decem, quorum quinque alterna Antheris carent *fig.* 7: ANTHERÆ ſaturate purpuraſcentes, *fig.* 6.

STAMINA: ten FILAMENTS, five of which want the Antheræ, the ANTHERÆ of a deep purple colour, *fig.* 6.

NECTARIA: *Glandulæ* quinque fuſcæ circa baſin ſtaminum locantur, *fig.* 9.

NECTARIA: five brown GLands placed round the baſe of the Stamina, *fig.* 9.

PISTILLUM: GERMEN quinquangulare, villoſum; STYLUS ſubulatus, ſulcatus; STIGMATA quinque purpuraſcentia, paululum reflexa, *fig.* 10, 11.

PISTILLUM: GERMEN quinquangular and villous, STYLE tapering and grooved; STIGMATA five, of a purple colour, bending a little back, *fig.* 10, 11.

PERICARPIUM nullum; FRUCTUS pentacoccus, roſtratus.

SEED-VESSEL none; FRUIT as yet unripe, formed of five protuberating ſeeds, and terminating in a long beak.

SEMEN oblongum, læve, fuſcum, arillatum, *fig.* 14, ARILLA hirſuta; ARISTA prælonga piloſa inſtructa quæ demum ſpiralis evadit, *fig.* 12, 13.

SEED oblong, ſmooth, brown, incloſed within an ARILLUS, *fig.* 14, which is hirſute, and furniſhed with a long hairy ARISTA, finally becoming ſpiral, *fig.* 12, 13.

We have often had occaſion to remark the very great difference in the appearance of plants ariſing from ſoil and ſituation; of this the young Botaniſt cannot be too well appriſed, nor too often informed: from a want of attention to this circumſtance, the plant which we have now deſcribed, has been divided by different Authors into ſeveral ſpecies.

It ſeems worthy of notice, that the alterations which are produced in plants from growing in a richer ſoil, are chiefly thoſe of encreaſe of ſize, and a multiplication of their parts; the minutiæ of the fructification ſuffer but little change in their form by culture, hence they are often moſt to be depended on, even in aſcertaining different ſpecies.

When the *Geranium Cicutarium* grows on a dry ſandy bank, or wall, as it very frequently does, it is quite diminutive; when it occurs in a moiſter and more luxuriant ſoil, the branches extend often a foot or two in length, and the whole plant becomes ſo altered in its general appearance, as readily to deceive the inexperienc'd Tyro; but the long pointed fruit which occurs in both, and from whence this plant has obtained the name of *Cranes-bill*, readily points them out to be the ſame.

The ſeeds of the Geraniums are, in general, encloſed within a membranous *Arillus*, which terminates in an *Ariſta* or *Tail*, of different lengths in different ſpecies; in ſome of them, when the ſeeds are become ripe, they detach themſelves from the receptacle, to which they are affixed, with conſiderable elaſticity, and the ſeeds being looſely contained within the *Arillus*, are thrown out to a conſiderable diſtance. In the preſent ſpecies, the ſeeds are more cloſely inveſted by the *Arillus*, which does not ſeparate itſelf with ſo much force, and as ſoon as detached, the *Ariſta* begins to be twiſted up in a ſpiral form. This may be very diſtinctly obſerved if we ſeparate a ſeed, with its *Arillus*, as ſoon as ripe, and place it in the palm of the hand, the tail of the *Arillus* immediately appears in motion, as if endued with ſome ſenſitive property, and continues uninterruptedly this motion, 'till it has aſſumed the form of a ſcrew, *vid. fig.* 13. The ſeed thus furniſhed with its twiſted Ariſta, is more liable to attach itſelf to any thing which may come in contact with it, by which means this plant is more univerſally diſſeminated.

The *Geranium moſchatum* has a great affinity with this ſpecies, that plant however has a ſtrong ſmell of muſk, which this entirely wants; and has alſo many other peculiarities, which we ſhall not fail to particularize when it comes to be treated of.

Geranium cicutarium.

GERANIUM ROBERTIANUM. STRONG-SCENTED CRANES-BILL, OR HERB ROBERT.

GERANIUM *Linnæi Gen. Pl.* MONADELPHIA DECANDRIA.

Stigmata quinque. *Fructus* rostratus, pentacoccus.

Raii Syn. 335. HERBÆ PENTAPETALÆ VASCULIFERÆ.

GERANIUM *robertianum* pedunculis bifloris, calycibus pilosis decemangulatis. *Linnæi Syst. Vegetab. p.*

515. *Fl. Suecic.* 241. *n.* 619.

GERANIUM foliis duplicato pinnatis, pinnis ultimis confluentibus, calycibus striatis, hirsutis. *Haller*

hist. n. 943.

GERANIUM *robertianum. Scopoli Fl. Carniol. n.* 845. *Hudson Fl. Angl. p.* 264.

GERANIUM *robertianum* primum. *Bauhin. Pin.* 319.

GERANIUM *robertianum. Gerard. emac.* 939.

GERANIUM *robertianum* vulgare. *Parkinson* 710. *Raii Syn. p.* 359.

RADIX annua, fusca, fibris ramosis prælongis instructa.

CAULES plures, diffusi, ramosi, sanguinei ut ut tota planta haud infrequenter, geniculis tumidis, pilosi, præsertim in junioribus plantis.

FOLIA opposita, pilosa, præcipue in umbrosis, unumquodque folium e tribus foliolis pinnatifidis basi confluentibus componitur, foliolo medio longius pedicellato, laciniis spinula rubra terminatis.

STIPULÆ ad singulum geniculum quatuor, utrinque binæ.

PEDUNCULI biflori.

CALYX: PERIANTHIUM decemangulatum, persistens, foliolis ovato-lanceolatis, nervosis, hirsutis, mucronatis, *fig.* 1, 2.

COROLLA : PETALA quinque rosea, patentia, æqualia, lamina subcordata, unguis linearis, medio prominulo sulcato in tres nervos albidos divaricante. *fig.* 3.

STAMINA : FILAMENTA decem fertilia, subulata, plana, alba, basi cohærentia ; ANTHERÆ purpurascentes, polline flavo repletæ, *fig.* 4, auct 5.

PISTILLUM : GERMEN quinquangulare ; STYLUS subulatus, villosus ; STIGMATA quinque, rubra, paululum reflexa, *fig.* 6.

SEMINA quinque Arillata, lævia, ovata, fusca ad unum latus compressa, *fig.* 9 ; ARILLUS rugosus, *fig.* 7, 8.

ROOT annual, brown, furnished with long branched fibres.

STALKS several, spreading, branched, of a blood-red colour, as is frequently the whole plant, (the joints turnid,) hairy, particularly in the young plants.

LEAVES opposite, hairy especially when growing in the shade, each composed of three pinnatifid leaves, uniting at the base, the middle leaf standing on the longest foot-stalk, the laciniæ or jags of the leaf terminated by a small red spine.

STIPULÆ four at each joint, two on each side of it.

PEDUNCLES biflorous.

CALYX : a PERIANTHIUM having ten angles, and continuing, the leaves ovato-lanceolate, nervous, hairy, terminating in a point, *fig.* 1, 2.

COROLLA: five rose-coloured PETALS, spreading and equal, the lamina somewhat heart-shaped, the claw linear, the middle part of it prominent, grooved, and spreading into three whitish nerves.

STAMINA : ten fertile FILAMENTS, tapering, flat, white, connected at bottom ; ANTHERÆ purplish, filled with a yellow Pollen, *fig.* 4, magnified, *fig.* 5.

PISTILLUM : GERMEN having five angles ; STYLE tapering, villous ; STIGMATA five, red, a little turned back, *fig.* 6.

SEEDS five, contained within an Arillus, smooth, oval, brown, flattened on one side, *fig.* 9 ; the ARILLUS wrinkled, *fig.* 7, 8.

Although our English *Geraniums* cannot boast that grandeur and variety of splendid colours so conspicuous in many of the foreign ones, yet several of them are sufficiently beautiful to be entitled to a place in the gardens of the curious, particularly the *Bloody Cranes-bill,* (*Geranium Sanguineum* ;) the *Crowsfoot Cranes-bill,* (*Geranium Pratense* ;) the *Perennial Doves-foot Cranes-bill,* (*Geranium Pereune of Hudson,*) and the *Herb Robert* which we have now described· the latter of these grows naturally in woods, but more particularly under the hedges which surround woods ; it likewise is frequently found in old hollow trees, and not uncommonly on the roofs of houses not much exposed to the sun : it is an annual plant ; the seeds sow themselves in Autumn, soon after the young plants come up ; flower the ensuing spring, and continue to blossom the whole Summer long, if the plant grows in the shade ; towards the latter end of the year, both stalks and leaves become of a deep red or blood colour.

The whole plant has a disagreeable smell when bruised, by which it will be distinguished from our other species. It appears to grow all over Europe, and as a proof of its being still more universal, LINNÆUS mentions its growing in *Arabia fælix.*

A variety with a white flower now and then occurs.

If credit may be given to writers on the *Materia Medica,* it is a plant of considerable efficacy in medicine, particularly as an Astringent, hence it is recommended in all kinds of Hemorrhages ; and those who have the management of cattle, are said to give them an infusion of this plant when they make bloody urine.—Has not this practice originated from the doctrine of signatures? It is also celebrated as a vulnerary in schrophulous, cancerous and putrid Ulcers, to which either the juice is applied, or the parts fomented with a decoction of the herb ; as likewise in Contusions, dissolving the extravasated blood when applied in the form of a Cataplasm ; and lastly it is said to be exhibited with good success in the Stone and Gravel.—How far it merits these encomiums future experiments must determine.

The herb bruised and applied to places infested with Bugs, is said by LINNÆUS to drive them away.

Géranium robertianum.

OROBUS *Linnæi Gen. Pl.* DIADELPHIA DECANDRIA.

Raii Synop. Gen. 23. HERBÆ FLORE PAPILIONACEO, SEU LEGUMINOSÆ.

OROBUS *tuberosus* foliis pinnatis, lanceolatis; ftipulis femifagittatis integerrimis, caule fimplici. *Lin. Syſt. Vegetab. p. 550. Fl. Suecic. n.* 642.

OROBUS caule fimplici; foliis fenis ellipticis; radice tuberofa. *Haller. hiſt. n.* 417.

ASTRAGALUS fylvaticus, foliis oblongis glabris. *Bauhin. pin.* 351. *Gerard. emac.* 1237.

LATHYRUS fylveſtris lignofior. *Parkinſon,* 1072. *Raii Synop. p.* 324. Wood-Peafe, or Heath-Peafe. *Hudfon, Fl. Angl. p.* 274. *Scopoli. Fl. Carn. n.* 883.

RADIX perennis, tuberofa.

CAULIS fimplex, erectus, pedalis, alatus, fubtortuofus.

FOLIA pinnata, CIRRHO brevi recto terminata, Pinnarum paria duo, tria, elliptica, mucronata, glabra fubtus cærulefcentia.

STIPULÆ femifagittatæ, fæpe integræ, fæpius vero ad bafin hamatæ, dente unico aut pluribus.

RAMI florigeri, 1, 2, 3, aut plures ex foliorum alis, primum nutantes, FLORES pulchelli, ex rubro purpurei, demum cærulefcentes.

CALYX PERIANTHIUM monophyllum, tubulatum, purpureum, bafi obtufum; ore quinquedentato, denticulis tribus inferioribus acutioribus, duobus fuperioribus brevioribus, obtufe divifis, fubafurgentibus, *fig.* 1.

COROLLA *Papilionacea:* VEXILLUM obcordatum, reflexum. *fig.* 2. ALÆ conniventes, Carinâ connexæ, Unguis linearis, *fig.* 5. Lamina obtufa. CARINA, *fig.* 7, acuminata, affurgens, marginibus cavis ad Alas recipiendas, *fig.* 9.

STAMINA: FILAMENTA diadelphia (fimplex et novem fidum) adfcendentia, *fig.* 11, 17. ANTHERÆ flavæ, *fig.* 12. ad bafin filamenti fimplicis et fuperioris, foramina duo obfervantur, *fig.* 16.

PISTILLUM: GERMEN cylindraceum, compreffum, STYLUS filiformis, erectus, lateri interiori prope apicem villofus, *fig.* 13.

PERICARPIUM LEGUMEN teres, longum, primum rubrum, demum nigrum, *fig.* 14.

SEMINA plura, fubrotunda, e luteo-fufca, *fig.* 15.

ROOT perennial and tuberous.

STALK fimple, upright, about a foot high, winged and fomewhat twifted.

LEAVES pinnated, terminated by a fhort ftrait CIRRHUS confifting of two or three pair of Pinnæ which are elliptical, and end in a fmall fharp point, fmooth and underneath blueifh.

STIPULÆ femifagittate, frequently entire but more often jagged at bottom, with one or feveral teeth.

BRANCHES which fuftain the flowers 1, 2, 3, or more, fpringing from the bofom of the leaves, at firft drooping · the FLOWERS beautiful, of a reddifh purple colour, becoming blue as they go off.

CALYX: a PERIANTHIUM of one leaf, tubular, purple, blunt at bottom, the mouth quinquedentate, the three lowermoft teeth fharpeft, the two uppermoft fhorteft, bluntly divided, and turned a little upwards, *fig* 1.

COROLLA *Papilionaceous:* the VEXILLUM heart-fhaped, turning back, *fig.* 2. the WINGS connivent and connected with the Carina, the Claw linear, *fig.* 5. the Lamina obtufe, *fig.* 6. the CARINA or Keel acuminate, rifing upward, the edges hollow for the reception of the Alæ or Wings, *fig.* 9.

STAMINA: ten FILAMENTS, nine united into one body below, and one feparate at top, *fig.* 11, 17. rifing upward, ANTHERÆ yellow, *fig.* 12. at the bafe of the fimple and uppermoft filament two fmall holes are confpicuous, *fig.* 16.

PISTILLUM: GERMEN cylindrical, and flattifh, STYLE thread-fhaped, interiorly near the tip villous, *fig.* 13.

SEED-VESSEL, a LEGUMEN round, and long, firft red, when ripe black, *fig.* 14.

SEEDS feveral, roundifh, of a yellowifh brown colour, *fig.* 15.

This elegant fpecies of Orobus grows very plentifully in all our Woods about Town; it feems to delight in a ftrong clayey foil. It produces its bloffoms in May and June and the feed is ripe in July. The root is large and tuberous, deeply fituated in the Earth and taken up with difficulty; it is not made any particular ufe of with us, but is confiderably efteemed in fome parts of Great Britain:

My very worthy and ingenious Friend the *Rev. Mr. Lightfoot,* of *Uxbridge,* has favoured me with the following account of its ufes, which he obferved in his late tour through Scotland:

"The *Orobus tuberofus* is very common in *Scotland,* both in the *Lowlands, Highlands,* and the *Hebrides.* It is called
"in the Erfe Language *Cor-meille.* The Highlanders dig up the Roots and dry them in their pockets, and chew
"them like Tobacco or Liquorice Root, to relifh their Liquor, and to repel Hunger and Thirft. In *Breadalbane*
"and *Rofs-fhire* they fometimes fteep them in Water, and make an agreeable fermented Liquor with them, which
"they efteem to be good for Diforders of the Thorax. It has a fweetifh Tafte fomewhat like Liquorice Roots. Fond
"as the Highlanders were of this Root they frequently ufed to change it with me for fome Pig-tail Tobacco, their
"favourite Indulgence."

Orobus tuberosus

Ervum hirsutum.

1   2   3   4   5   6   7

Ervum hirsutum. Rough-podded Tine-Tare.

ERVUM *Linnæi Gen. Pl.* Diadelphia Decandria. Calyx quinquepartitus, longitudine corollæ.

Raii Gen. 23. Herbæ flore papilionaceo seu leguminosæ.

ERVUM *hirsutum*, pedunculis multifloris, seminibus globosis binis. *Linn. Syst. Vegetab. p.* 554. *Spec. Plant.* 1039. *Fl. Suecic.* 255.

VICIA foliis linearibus, siliquis racemosis, dispermis, hirsutis. *Haller hist. helv. n.* 422.

ERVUM *hirsutum. Scopoli Fl. Carniol n.* 901. *Hudson Fl. Angl. p.* 280.

VICIA segetum cum siliquis plurimis hirsutis. *Baubin. Pin. p.* 345.

VICIA sylvestris seu Cracca minima. *Gerard. emac.* 1028.

ARACHUS sive Cracca minor. *Parkinson* 1070. *Raii Syn.* small wild Tare or Tine Tare. *Muller. Flor. Dan. icon.* 639.

RADIX annua, tenuis, prælonga, paucis fibrillis instructa. *fig. 1.*

CAULES pedales, aut bipedales, debiles, ramosi, quadrangulares, tortuosi.

STIPULÆ in plures lacinias tenues divisæ, superiore majore.

FOLIA pinnata, ad octo aut duodecem paria, opposita, aut subalterna, lævia, lanceolata, *apice truncato, nervo medio in mucronem educto,* capreolo ramoso terminata.

PEDUNCULI longitudine foliorum, *multiflori.*

FLORES a tribus ad octo, pallide purpurei, racematim, et imbricatim dispositi.

CALYX: Perianthium quinquedentatum, persistens, longitudine fere Corollæ, dentibus linearibus, acuminatis, subæqualibus, duobus superioribus more Orobi obtuse divisis, *fig.* 1.

COROLLA papilionacea; Vexillum subrotundum, vix emarginatum, parum reflexum, *fig.* 2; Alæ Carinæ adhærentes, ovatæ, obtusæ, ad basin lineares, *fig.* 3; Carina alis brevior, *fig.* 4, *interne maculâ purpureâ utrinque notata.*

STAMINA: Filamenta decem assurgentia, supremum brevior cæteris, nec liberum, *fig.* 5; Antheræ simplices, flavæ.

PISTILLUM: Germen oblongum, Stylus simplex, assurgens, Stigma obtusum, villosum, *fig.* 6.

PERICARPIUM: Legumen breve, *hirsutum, dispermum, fig.* 7.

SEMINA duo, subrotunda.

ROOT annual, slender, long, and furnished with few fibres.

STALKS from one to two feet high, weak, branched, quadrangular and twisted.

STIPULÆ divided into many slender laciniæ, of which the uppermost is the largest.

LEAVES pinnated, from eight to twelve pair, opposite, or nearly alternate, smooth lanceolate, with the *top cut off,* and the *midrib running out to a short point,* terminated by a branched tendril.

PEDUNCLES the length of the leaves, and supporting many flowers.

FLOWERS from three to eight, of a pale purple colour, disposed in racemi, and laying one over another.

CALYX: a Perianthium with five teeth, continuing, almost the length of the Corolla, the teeth linear, and pointed, nearly equal, the two upper ones obtusely divided in the manner of the Orobus, *fig.* 1.

COROLLA papilionaceous; the Vexillum roundish, scarcely nicked in, bending a little back, *fig.* 2; the Wings adhering to the Carina, oval, obtuse, at bottom linear, *fig.* 3; the Carina shorter than the Wings, *fig.* 4, *marked internally on each side with a purple spot.*

STAMINA: ten Filaments which rise upward, the uppermost connected with, and shorter than the others, *fig.* 5; the Antheræ simple and yellow.

PISTILLUM: Germen oblong, Style simple and rising upward, Stigma blunt and villous, *fig.* 6.

SEED-VESSEL a short *hairy* Legumen with *two seeds, fig.* 7.

SEEDS two, and roundish.

This species of Tine-Tare, which at first sight bears so great a resemblance to the *Ervum tetraspermum,* grows like that, too frequently among Corn, to which it is in general more destructive, as being a stronger and more prolific plant. I have in wet seasons seen whole fields of corn overpower'd and wholly destroyed by this plant.

It is easily distinguished from the *Tetraspermum;* in the first place, the leaves are not pointed as in that species, but appear as if cut off at the end, which although a material circumstance is not noticed by Muller in his figure of it, *vid. Fl. Dan. icon.* 639; secondly the Stipulæ are divided into many more laciniæ; the flowers and consequently the Pods grow in a kind of Cluster, whereas there is seldom more than two grow together in the *Tetraspermum;* and lastly, which seems to be the best distinction, the Pods are rough and contain two Seeds in each, while in the *Tetraspermum,* they are smooth and contain four Seeds.

ERVUM *Linnæi Gen. Pl.* DIADELPHIA DECANDRIA.

Raii Syn. Gen. 23. HERBÆ FLORE PAPILIONACEO SEU LEGUMINOSÆ.

ERVUM *(tetraspermum)* pedunculis subbifloris, seminibus globosis quaternis. *Linn. Syst. Vegetab. p.* 554.

VICIA foliis linearibus, filiquis gemellis glabris. *Haller hist. v.* 1. *p.* 184.

ERVUM *tetraspermum. Scopoli Fl. Carniol.* DIAGN. Pedunculi fubbiflori. Siliqua glabra, obtufa, tetrafperma.

VICIA fegetum fingularibus filiquis glabris. *Bauhin Pin. p.* 345.

VICIÆ five Craccæ minimæ fpecies cum filiquis glabris. *I. Bauhin.*

CRACCA minor filiquis fingularibus, flofculis cærulefcentibus. *Hoff. C. H. Alt. Raii Syn. p.* 322. Tine-

Tare with fmooth pods. *Hudfon Fl. Angl p.* 280. *OEder Fl. Dan. Icon.* 95.

RADIX annua, fibrofa.	ROOT annual and fibrous.
CAULES in apertis locis læves, tenues, debiles, inter fegetes vero, (ubi fæpius invenitur) capreolis erecte fefe fuftentant, pedales et ultra.	STALKS in open places are flender and weak, but among the Corn, (where this plant is moft commonly found,) they fupport themfelves upright by means of their tendrils, and grow to a foot or more in height.
STIPULÆ ad bafin foliorum, duo, fimplices, utrinque acuminatæ.	STIPULÆ at the bottom of the leaves, two, fimple, and pointed at each end.
FOLIA pinnata, lævia, lanceolata-linearia, parium trium ad quinque ufque, capreolo ramofo terminata.	LEAVES pinnated, fmooth, lanceolate and linear, from three to five pair, terminated by a branched tendril.
PEDUNCULI longitudine foliorum, plerumque biflori.	PEDUNCLES the length of the leaves, generally fuftaining two flowers.
CALYX PERIANTHIUM quinquedentatum, perfiftens, dentibus inæqualibus, acutis, duobus fuperioribus brevioribus, latioribus, furfum tendentibus, obtufe divifis, *fig.* 1.	CALYX a PERIANTHIUM having five teeth and continuing, the teeth unequal and pointed, the two uppermoft fhorteft, broadeft, and turning a little upwards, at bottom obtufely divided, *fig.* 1.
COROLLA papilionacea, *fig.* 2; VEXILLUM fubemarginatum, limbus reflexos, venis purpureis pictus, *fig.* 4; Alæ albæ, conniventes, *fig.* 5; CARINA alis brevior, obtufa, *fig.* 6.	COROLLA papilionaceous, *fig.* 2; the VEXILLUM flightly nicked in at top, the limb fomewhat turned back and ftreaked with purple. *fig.* 4; the ALÆ white and clofing together, *fig.* 5; the CARINA fhorter than the Alæ and obtufe *fig.* 6.
STAMINA: FILAMENTA diadelpha (fimplex et novemfidum) affurgentia, *fig.* 7, 8, fupremum liberum, *fig.* 8; ANTHERÆ fimplices.	STAMINA: Ten FILAMENTS uniting into two bodies, of which one forms the lowermoft, *fig.* 7, and one the uppermoft which is free, *fig.* 8; ANTHERÆ fimple.
PISTILLUM: GERMEN compreffum; STYLUS affurgens; STIGMA capitatum, villofum, *fig.* 9.	PISTILLUM: GERMEN flatten'd; STYLE rifing upward; STIGMA forming a little head and villous, *fig.* 9.
PERICARPIUM: LEGUMEN læve, teretiufculum, tetrafpermum, *fig.* 10.	SEED-VESSEL: a LEGUMEN, fmooth, roundifh, and containing four feeds, *fig.* 10.
SEMINA fubrotunda, fufcefcentia, nigro marmoreata, *fig.* 11.	SEEDS nearly round, brownifh and mottled with black, *fig.* 11.

This fpecies of *Ervum* or *Tine-Tare* is found in moft Corn-fields, often to the Farmers forrow, as it frequently proves very injurious to the Corn, laying hold of it by means of its tendrils, and if the feafon favours its growth quite overcoming it. Like moft plants of this kind it is exceedingly fertile; on one plant which I cafually pulled up, I counted 220 pods, and as each pod contains four feeds, there muft have been from a fingle feed the amazing produce of 880.

At firft fight this fpecies has a confiderable refemblance to the *Ervum hirfutum*, but the flighteft attention will difcover the difference; in the *Ervum hirfutum* the pods contain only *two feeds* and are *hairy*; in the *Tetrafpermum* they contain *four* and are *fmooth*; in the *hirfutum* the flowers grow in a kind of clufter, in *this fpecies* there is feldom more than *two* grow together.

The figure which I have given is intended to reprefent the plant as it grows among the Corn; when it is found by itfelf and in a poor foil it is often not fo large.

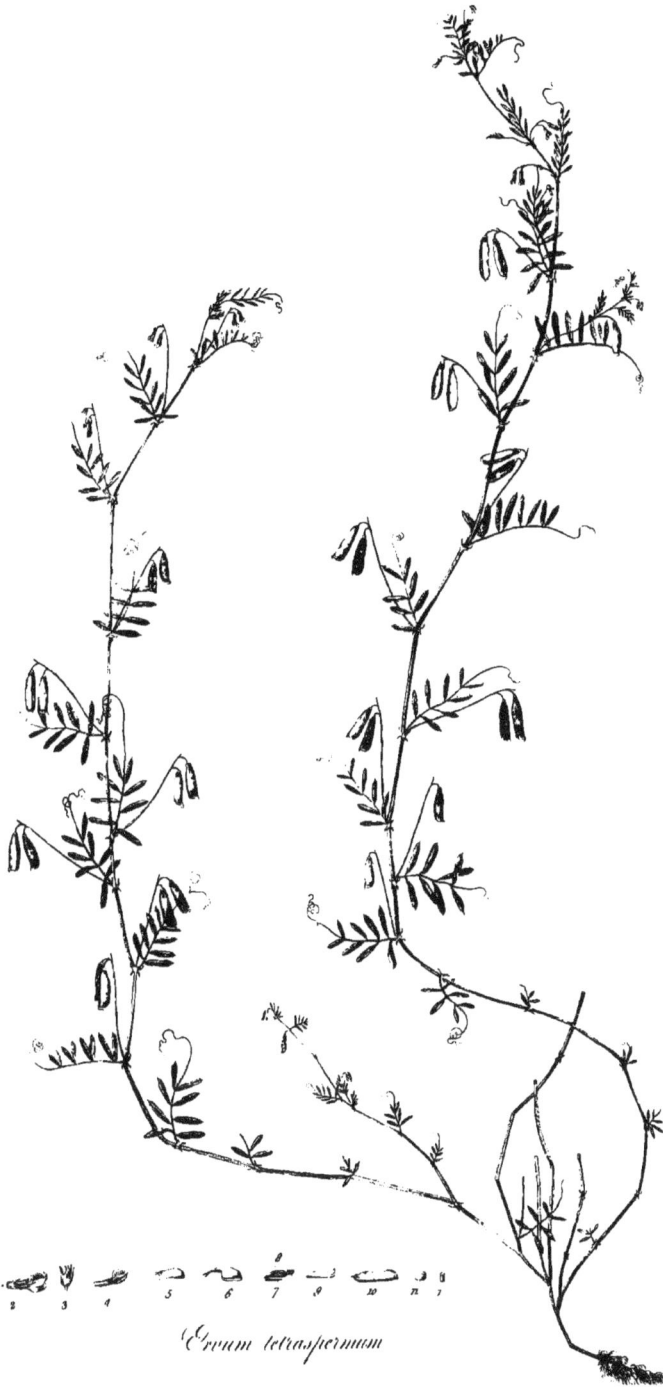

Ervum tetraspermum

Hypericum pulchrum

HYPERICUM PULCHRUM. Small Upright St. JOHN's WORT.

HYPERICUM *Linnæi Gen. Pl.* POLYADELPHIA POLYANDRIA.

Raii Synop. Gen. 24. HERBÆ PENTAPETALÆ VASCULIFERÆ.

HYPERICUM floribus trigynis ; calycibus ferrato-glandulofis, caule tereti, foliis perfoliatis glabris. *Lin. Sp. Pl.* 1106.

HYPERICUM pulchrum Tragi. *J. Bauhin. Hift.* III. 183. *Raii Synop.* 342.

HYPERICUM minus, erectum. *Bauhin. Pin.* 279.

HYPERICUM foliis amplexicaulibus, cordatis, calycibus ovatis, ferratis, glanduligeris. *Haller. Hift. n.* 1041.

Gerard. emac. 540. *Hudfon. Fl. Angl.* 290. *Order. Flor. Dan. Icon.* 75.

RADIX perennis.

CAULIS pedalis ad bipedalem, erectus, *teres*, *fig.* 1. glaber, fubramofus, geniculi diftantes.

RAMI oppofiti, breves, tenues, cauli fimiles.

PEDUNCULI teretes, plerumque triflori.

FOLIA CAULIS *cordato-triangularia, glaberrima, amplexicaulia*, faturate viridia, patentia, quam in cæteris Hypericis folidiora, verfus marginem perforata, inferiora frequenter coccinea, RAMORUM ovata, caulis triplo minora, PEDUNCULORUM ovato-lanceolata.

CALYX. PERIANTHIUM quinquepartitum, LACINIIS ovatis, acutis, ftriatis, margine ferratis, *dentibus glanduliferis*, glandulis nigro rufis, *fig.* 2.

COROLLA. PETALA quinque, oblongo-ovata, flava, contorta, leviter ftriata, fubtus aurantiaco lineata, margine fubferrata et glandulis cincta, *fig.* 3.

STAMINA. FILAMENTA triginta fex, filiformia, in tres fafciculos ad bafin coalita, in fingulo fafciculo duodecim ; ANTHERÆ biloculares, fubrotundæ. POLLEN *miniaceum, fig.* 4.

PISTILLUM. GERMEN ovatum, STYLI tres, longitudine germinis, divaricantes, STIGMATA parva, fubrotunda, *fig.* 5.

PERICARPIUM. CAPSULA fubconica, trilocularis, fufca, *fig.* 6, 7.

SEMINA plurima, oblonga, fufca, *fig.* 8.

ROOT perennial.

STALK from one to two feet high, upright, round, *fig.* 1. fmooth, and thinly branched, the joints remote from each other.

BRANCHES oppofite, fhort, flender and like the ftalk.

PEDUNCLES round, generally fuftaining three flowers.

LEAVES of the STALK *triangularly heart-fhaped, fmooth, fhining, embracing the ftalk,* nearly horizontal, of a deep green colour, more folid to the touch than the other St. John's Worts, perforated near the edge, and frequently of a bright red colour towards the bottom, thofe of the BRANCHES oval, three times fmaller than thofe on the Stalk, and thofe of the PEDUNCLES lancet-fhaped.

CALYX. A PERIANTHIUM divided into five Segments, the SEGMENTS oval, pointed, ftriated, ferrated and, *edged with little glands* of a blackifh red colour, *fig.* 2.

COROLLA. Five PETALS, oblong, oval, yellow, flightly ftriated ; on the under fide tinged with a bright orange, flightly ferrated, and edged with glands, *fig.* 3.

STAMINA. The FILAMENTS numerous, to thirty-fix, filiform, uniting at bottom in three Fafciculi or Bundles, in each Fafciculus 12 ; the ANTHERÆ roundifh and bilocular, *fig.* 4. The POLLEN bright fcarlet.

PISTILLUM. The GERMEN oval, three STYLES the length of the Germen, fpreading, the STIGMATA fmall and roundifh, *fig.* 5.

SEED-VESSELL a CAPSULE fomewhat conical, of a brown colour, with three cavities, *fig.* 6, 7.

SEEDS numerous, oblong and brown, *fig.* 8.

THE antient Botanifts gave this Plant the name of PULCHRUM, from its beauty ; and LINNÆUS has very properly continued it ;—many will, no doubt, think it deferving of a place in their gardens, it is fond of a clayey foil, and woody fituation, and is found in all the woods about town, as HORNSEY WOOD beyond *Iflington*, OAK OF HONOUR WOOD, (as it is generally called) a little beyond *Peckham* ; CHARLTON-WOOD by *Greenwich*, likewife on HOUNSLOW-HEATH ; it flowers in the month of *July*, and continues but a fhort time in bloffom.

Its virtues as a medicine are probably the fame with the common *St. John's Wort.*

Hypericum Perforatum

HYPERICUM PERFORATUM. COMMON St. JOHN's WORT.

HYPERICUM   *Linnæi Gen. Pl.* POLYADELPHIA POLYANDRIA.

       Raii Synopfis Gen. 24. HERBÆ PENTAPETALÆ VASCULIFERÆ.

HYPERICUM   *perforatum,* floribus trigynis, caule ancipiti, foliis obtufis pellucido-punctatis. *Linnæi Syft.*
       Vegetab. p. 584. *Fl. Suecic. n.* 680.

HYPERICUM   caule terete, alato, ramofiffimo; foliis ovatis, perforatis. *Haller. hift. vol.* 2. *p.* 4.

HYPERICUM   vulgare *Bauhin. Pin. p.* 279. *Gerard. emac.* 540. *Parkinfon* 572. *Raii Synop.* 342. *Hudfon Fl.*
       Angl. 290. *Scopoli. Fl. Carniol. n.* 944.

Tota planta glandulis nigris adfperfa.	The whole plant is fprinkled over with fmall black glands.
RADIX perennis, lignofa, fufca.	ROOT perennial, woody, of a brown colour.
CAULES plerumque plures ex eadem radice, bipedales, erecti, fubliguofi, *læves, teretes, alternè ancipites fig.* 1, ramofi.	STALKS feveral for the moft part, fpringing from the fame root, about two feet high, upright, woody, *fmooth, round, alternately two edged, fig.* 1, much branched.
RAMI oppofiti, fuberecti, ancipites.	BRANCHES oppofite, nearly upright, two edged.
FOLIA oppofita, feffilia, ovato-oblonga, obtufa, perforata five pellucido-punctata, heptanervia ex luteo-viridia. *fig.* 2.	LEAVES oppofite, feffile, of an oblong oval fhape, obtufe, having the appearance of being all over perforated, of a yellowifh green colour, with feven nerves or ribs, *fig.* 2
PEDUNCULI ancipites, multiflori.	PEDUNCLES two edged, fupporting many flowers.
PANICULA denfa.	PANICLE bufhy.
CALYX: PERIANTHIUM quinquepartitum, ftriatum, laciniis *lanceolatis, acuminatis, nudis. fig.* 3.	CALYX A PERIANTHIUM divided into five fegments, and ftriated, the fegments *narrow and pointed, without any glands on them. fig.* 3.
COROLLA: petala quinque, flava, ad unum latus crenulata, glandulis nigris adfperfa. *fig.* 4.	COROLLA: five PETALS of a yellow colour, notched irregularly on one fide, and fprinkled over with little black glands. *fig.* 4.
STAMINA: FILAMENTA plurima, in tria corpora vix coalita. *fig.* 5. ANTHERÆ flavæ, biloculares, loculis fubrotundis, inter quos *glandula nigra* ponitur. *fig.* 6.	STAMINA: FILAMENTS numerous, uniting at bottom in three fcarcely diftinct bodies or fafciculi *fig.* 5. ANTHERÆ yellow and bilocular, each of the Cavities of a roundifh figure, and between them is fituated *a fmall black gland. fig.* 6.
PISTILLUM: GERMEN fubovatum, STYLI tres divaricantes: STIGMATA fimplicia. *fig.* 7.	PISTILLUM: GERMEN fomewhat oval, three STYLES which divaricate; the STIGMATA fimple, *fig.* 7.
PERICARPIUM: CAPSULA fubtrigona *fig.* 8. trilocularis *fig.* 9. pallide fufca.	SEED-VESSEL: a CAPSULE fomewhat triangular, *fig.* 8, of a pale brown colour, with three Cavities, *fig.* 9.
RECEPTACULUM feu Thalamus feminum foramine triquetro gaudet, quod in pericarpii immaturi fectione tranfverfa clare diftingui poteft, ut obfervavit Cl. Scopoli.	RECEPTACLE: the Receptacle which is continued through the Capfule, and connects the Cavities together, has a triangular hole in it, which is very obvious in a tranfverfe fection of it before it is ripe,—as the celebrated *Scopoli* has juftly obferved.
SEMINA plurima, oblonga, fufca. *fig.* 10. 11.	SEEDS numerous, oblong, and brown, *fig.* 10. 11.

It very often happens, that fome of the minute parts of the Flower, and Seed, afford a more obvious, certain, and conftant mark of fpecific difference, than any part of the plant befides, and we have a remarkable inftance of the truth of this obfervation in the plant before us. A little gland, of a black colour, placed on the fummit of the Anthera, at one view diftinguifhes this fpecies, without any farther inveftigation: did fuch obvious diftinctions prevail in all plants, a knowledge of them might with much eafe be acquired; and fortunately we fhall find, on examination, fuch marks more frequently occur than is generally imagined; whenever they do, we fhall not fail to remark them.

The apparent perforation of the leaves, from whence this fpecies is named, is not peculiar to it alone.

Although in the prefent practice this officinal plant does not feem to be much regarded, yet its fenfible qualities, and the repeated teftimonies of its virtues, entitle it as Dr. Cullen * obferves to farther trials. To the tafte it is aftringent and bitter, and its effects feem to be chiefly diuretic. From poffeffing properties which have generally been called balfamic, it has been ufed as a vulnerary in external wounds, and internal hæmorrhages, for the former purpofe, the tops of the plant with the flowers are infufed in oil, and for the latter, an infufion of the plant is made in the manner of Tea. It has likewife been given in ulcerations of the kidnies, and has even been fuppofed to poffefs virtues as a febrifuge.

It has had the ill fate to be abufed by the fuperftition of the common people in France and Germany, who gather it with great ceremony on St. John's Day, and hang it in their Windows, as a certain charm and defence againft Storms, Thunder, and evil Spirits; miftaking the meaning of fome medical writers, who have fancifully given this plant the name of *Fuga Dæmonum* becaufe they fuppofed, if given internally, it was a good medicine for maniacal and hypochondriacal Diforders.

The dried plant boiled with Allum dyes Wool of a yellow colour. It grows very common in hedges and fields that are but feldom tilled, and flowers in Auguft and September.

*Vid. Dr. Cullen's Materia Medica p. 206.

Leontedon Taraxacum.

LEONTODON *TARAXACUM.* DANDELION.

LEONTODON   *Linnæi Gen. Pl.* SYNGENESIA. POLYGAMIA ÆQUALIS.

            Raii Synopsis ed. 3. *Gen.* 6. HERBÆ FLORE COMPOSITO, NATURÆ PLENO LACTESCENTES.

LEONTODON   *Taraxacum* calycis squamis inferne reflexis, foliis runcinatis denticulatis lævibus. *Linnæi Syst. Vegetab. p.* 596. *Sp. Plant* 1122. *Fl. Suecic.* 270.

TARAXACUM   calycibus glabris, squamis imis reflexis. *Haller hist. v.* 1. *p.* 56.

HEDYPNOIS   *Taraxacum Scopoli Flor. Carn. n.* 957.

HEDYPNOIS   major *Fuschii.*

DENS LEONIS   latiore folio *Bauhin. Pin. p.* 126. *Gerard. emac.* 290, *Parkinson* 780. *Raii Syn. ed.* 3. *p.* 170. *Hudson Fl. Angl. p.* 297. *Oeder. Fl. Dan. Icon.* 574.

RADIX perennis, subfusiformis, lactescens, externe pallide fusca.

ROOT perennial, tapering, milky, externally of a pale brown colour.

FOLIA laciniato-pinnatifida, plus aut minus profunde incisa, laciniis acutis et acute dentatis, plerumque lævia, nonnunquam vero subaspera.

LEAVES more or less deeply jagged, each jag or lacinia pointed, and sharply indented, generally smooth, but sometimes a little rough.

SCAPI nudi, fistulosi, lactescentes, versus apicem subtomentosi, uniflori.

STALKS naked, hollow, milky, towards the top covered with a kind of down, supporting one flower on each.

CALYX communis lævis, glaucus, *squamis inferioribus reflexis, fig.* 1.

CALYX : the *common* or general Calyx smooth, glaucous, *the lowermost leaves or squamæ turning back, fig.* 1.

COROLLA *composita,* flava, corollulis hermaphroditis, numerosis, æqualibus. *Propria* monopetala, ligulata, truncata quinquedentata, *fig.* 2.

COROLLA : the flower compounded of a great number of COROLLULÆ or lesser flowers, which are yellow, hermaphrodite and equal; each *Corollula* monopetalous, tubular at bottom, and flat towards the extremity, the apex truncated and quinquedentate. *fig.* 2.

STAMINA : FILAMENTA quinque capillaria, brevissima, *fig.* 3. ANTHERÆ flavæ, in tubum cylindraceum coalitæ, *fig.* 4.

STAMINA : five FILAMENTS small and very short, *fig.* 3. the ANTHERÆ yellow, uniting and forming a cylindrical tube. *fig.* 4.

PISTILLUM : GERMEN oblongum, *fig.* 5, STYLUS longitudine corollæ, *fig.* 6. STIGMATA duo revoluta, *fig.* 7.

PISTILLUM : GERMEN oblong, *fig.* 5. STYLE the length of the COROLLA, *fig.* 6. STIGMATA two, rolling back, *fig.* 7.

SEMEN subincurvatum, subcompressum, subtetragonum, striatum, *apice echinatum,* pallide oliváceum, *fig.* 8, 9. PAPPUS stipitatus, simplex, stipite brevior, *fig.* 10

SEED a little crooked, flattish, and somewhat four square, striated or grooved, *at top prickly,* of a pale olive colour, *fig.* 8, 9. the *Down* or pappus standing on a footstalk, simple, not feathery, shorter than the footstalk, *fig.* 10.

RECEPTACULUM nudum, alveolatum. *fig.* 11.

RECEPTACLE naked, and full of little holes, *fig.* 11.

As a medicinal plant the Dandelion is thought to possess considerable virtues, and has been frequently made use of in obstructions of the Viscera, particularly the Jaundice. Some recommend the juice, others a decoction of the whole plant. It appears to operate chiefly by urine, and from possessing this property in a considerable degree it has acquired its vulgar name of *Piss-a-bed.* Its other, and more common name, seems to be a corruption of the French term *Dent de Lion.*

As a kind of sallad, this plant is by many prefered to any other, particularly by the inhabitants of Spitalfields, many of whom being descended from French families, that forsook their native country for one more favourable to religious liberty, still retain the peculiar customs of that people in their diet, &c. They blanch, or whiten it as the gardeners do endive, and the inferior class generally use the simple process of laying a tile on it, for whatever excludes the light from this or any other plant will make it become white, all plants deriving their colours from the fountain of light, the sun. And it is remarkable, that many plants containing bitter and acrid juices are rendered by this process mild, sweet, and agreeable : who, for instance, could eat endive, celery, or even lettuce, in their wild uncultivated states ?

The Dandelion grows in the greatest plenty in rich meadows, although it is very common on walls, and in courts and areas ; when growing in a barren soil or dry situation the leaves become more narrow and jagged.

It flowers in May, and is the first plant which covers our meadows with a beautiful yellow coat, a few weeks afterwards, when it produceth its seed, it changes this for a white one.

Children frequently amuse themselves with blowing off the seeds, which stand naked on the receptacle or top of the stalk, and the round white heads, formed by the expansion of their pappus or down, they call *clocks.*

The young botanist generally finds some difficulty in acquiring a clear idea of the structure of these compound flowers, occasioned by the minuteness of the parts of fructification, which however are much larger and more conspicuous in this than in many others of the class SYNGENESIA, and therefore a proper flower for him to begin with.

On examining the flower of the Dandelion he will find that it is not a double flower, properly so called, as he might be led to think from its fullness, but that it is composed of a great number of *Flosculi,* or lesser flowers placed close together on one common receptacle or bottom, and enclosed by one common or general calyx. On dissecting each of these *Flosculi,* he will find them to consist of a COROLLA, or PETAL *fig.* 2, which at bottom is tubular, but towards the extremity flat, that from the bottom or tubular part of the corolla, five FILAMENTS spring, which are small and short, yet loose and unconnected *fig.* 3, that these filaments are furnished with ANTHERÆ, which unite together and form a long slender tube *fig.* 4, beneath the corolla is placed the GERMEN, or future seed *fig.* 5, from whence the STYLE, or middle part of the Pistillum proceeds and passes up through the middle of the flower, betwixt the Filaments and through the tube formed by the union of the Antheræ, *fig.* 6, and is furnished at top with two STIGMATA which roll back, *fig.* 7, at a little distance from the Germen the lower part of the Stylus is surrounded by numerous upright hairs which are the future PAPPUS or Down, *fig.* 10.

This, then, he will find to be the appearance of the parts of fructification in a full blown flower.

Those parts of the flower which were more immediately or more remotely necessary to the impregnation of the Seed having now performed their office decay, the Corolla with the Stamina and upper part of the Pistillum drops off, the Seed becomes larger, the lower part of the Pistillum remains, is elongated and becomes the footstalk of the Pappus, and the Seed as yet immature with the Pappus as yet moist are all enclosed and pressed by the Calyx into a conical form. This is its appearance in its second state.

The fructification still going forward the seed becomes ripe and brown, the PAPPUS now deprived of its moisture expands itself every way, *fig.* 10, pushes back the Calyx, and assumes a spherical form. The seeds fitted for vegetation and thus exposed are carried away by the first strong wind, and " a new race planted far from their native soil."

Such then is the curious process which nature makes use of in the perfecting and dissemination of this plant.

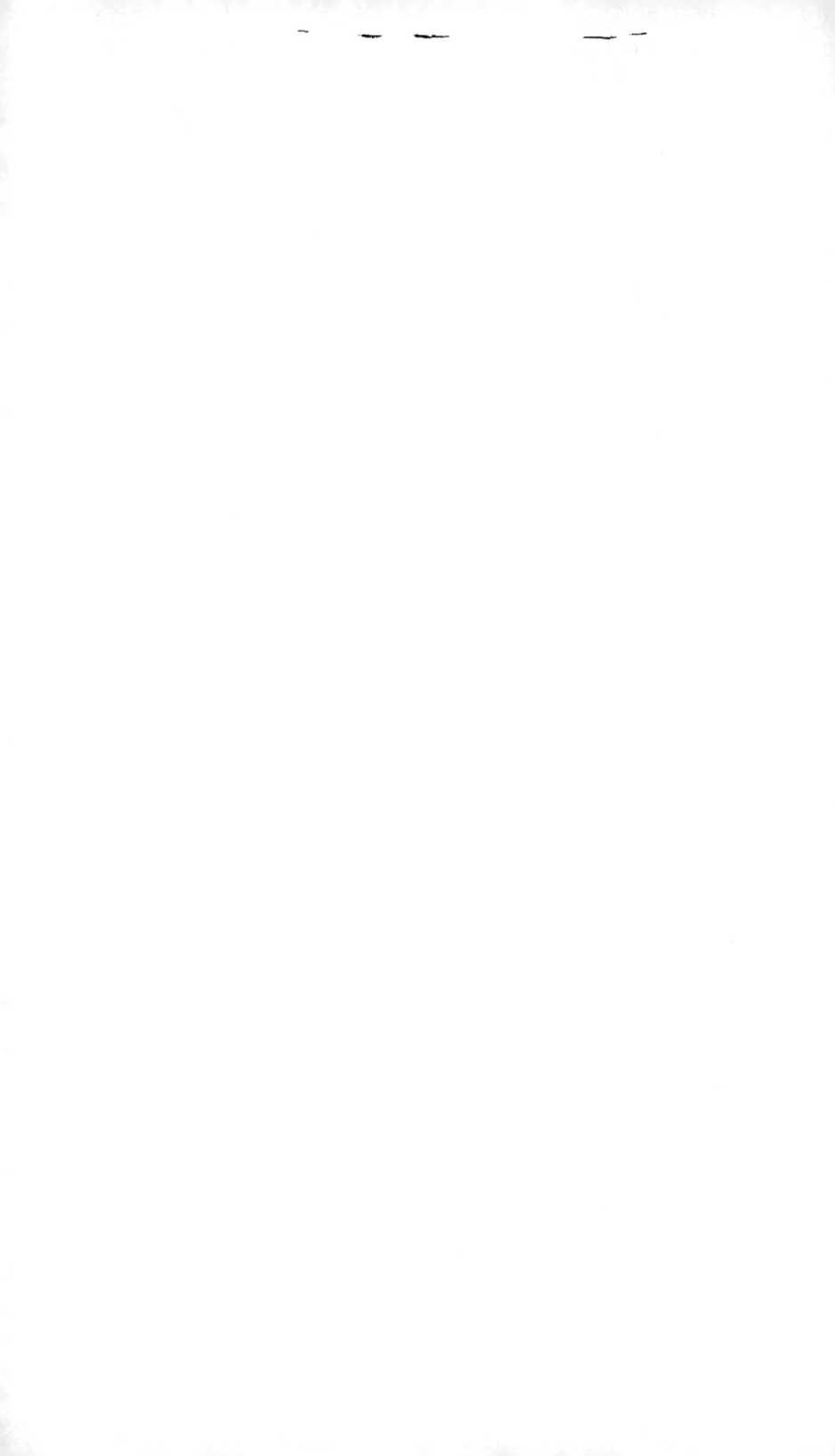

LAPSANA COMMUNIS. NIPPLEWORT.

LAPSANA *Linnæi Gen. Pl.* SYNGENESIA POLYGAMIA ÆQUALIS.

Receptaculum nudum. *Cal.* calyculatus, fquamis fingulis interioribus ca-
naliculatis.

Raii Syn. Gen. 6. HERBÆ FLORE COMPOSITO NATURA PLENO LASTESCENTES.

LAPSANA *communis* calycibus fructus angulatis pedunculis tenuibus ramofiffimis. *Linnæi Syft. Vegetab.*
p. 602. *Sp. pl.* 1141. *Fl. Suecic. p.* 277.

LAMPSANA caule brachiato ; foliis ovatis longe petiolatis ; petiolis pinnatis. *Haller hift. n.* 6.

LAMPSANA *communis. Scopoli Fl. Carniol. n.* 988.

SONCHO affinis Lampfana domeftica. *C. Bauhin pin. p.* 124.

LAMPSANA *Gerard. emac.* 255.

LAMPSANA vulgaris. *Parkinfon* 810. *Raii Syn.* 173. *Hudfon Fl. Angl. p.* 303.

RADIX annua, fimplex, fibrofa.	ROOT annual, fimple, and fibrous
CAULIS erectus, rigidus, bicubitalis, ftriatus, ramofus, hirfutus.	STALK upright, rigid, about two cubits high, ftriated, branched, hairy.
FOLIA oppofita, hirfutula, ad radicem et in ima parte caulis uno vel altero pinnularum pari donata, fegmento terminali magno, ovato, dentato, fuperiora oblonga, dentata.	LEAVES oppofite, fomewhat hairy, at the root, and on the lower part of the ftalk furnifhed with one or two pair of pinnulæ ; the fegment which terminates the leaf large, oval, and indented ; the upper leaves oblong and indented.
CALYX : communis calyculatus, angulatus, lævis, fquamæ ad bafin minimæ erectæ, *fig.* 1.	CALYX ; the common Calyx fmooth, and furnifhed at bottom with a few minute, upright fquamulæ, *fig.* 1.
COROLLA compofita, imbricata, Corollulis hermaphroditis æquabbus ; *propria* monopetala, ligulata, truncata quinque dentata, *fig.* 2.	COROLLA compound, imbricated, the flofcules hermaphrodite and equal ; each of them monopetalous, ligulate, truncated, and having five teeth, *fig.* 2.
STAMINA : FILAMENTA quinque capillaria, breviffima ; ANTHERÆ cylindracea tubulofa, *fig.* 2.	STAMINA : five fmall and very fhort FILAMENTS ; ANTHERÆ uniting into a tube, *fig.* 2.
PISTILLUM : GERMEN oblonginfculum ; STYLUS filiformis, longitudine Staminum ; STIGMA bifidum, reflexum, *fig.* 2.	PISTILLUM : GERMEN oblong ; STYLE filiform, the length of the Stamina : STIGMA bifid and turning back, *fig.* 2.
SEMINA circiter octodecem, oblonga paululum incurvata, pappo deftituta, intra calycem, *fig.* 3, 4.	SEEDS about eighteen, oblong, a little bent in, without any down, contained within the Calyx, *fig.* 3, 4.

In gardens as a weed, this plant anfwers very well to the name of *communis,* being in general too common. Nature feems amply to have fupplied the want of pappus or down in the feeds, by the great number of them produced in each plant. It alfo occurs on the fides of banks, and in all cultivated ground ; flowering during moft of the Summer months.

According to RAY it receives its name of *Nipplewort* from its efficacy in curing fore nipples : no other virtues or ufes feem attributed to it.

Lapsana communis.

Erigeron acrez?

ERIGERON ACRE. PURPLE ERIGERON.

ERIGERON *Linnæi. Gen. Pl.* SYNGENESIA POLYGAMIA SUPERFLUA.

    Raii Synopfis. HERBÆ FLORE COMPOSITO, SEMINE PAPPOSO NON LACTESCENTES, FLORE DISCOIDE.

ERIGERON *Acre* pedunculis alternis unifloris. *Lin. Sp. Pl.* 1211.

ERIGERON *polymorphum Scopoli. Fl. Carniol.* DIAGN. folia lanceolata, bafi et apice attenuata. Germina villofa. Pappus ruffus.

ERIGERON caule alterne ramofo, petiolis unifloris, femifiofculis pappum æquantibus, et femifiofculis pappum fuperantibus *Haller. hift. n.* 85. 86.

CONYZA cærulea acris *Bauhin Pin.* 265. *Gerard emac.* 484.

ASTER arvenfis cæruleus acris. *Raii Syn.* 175. Blue-flowered fweet Fleabane.

CONYZA odorata cærulea *Parkinfon* 126.

SENECIO five Erigeron cœruleus *I. B.* II. 1043 *Hudfon Fl. Angl.* 314. *Oeder Fl. Dan. Tab.* 292.

RADIX perennis, fibrofa, fibris pallide fufcis.

ROOT perennial and fibrous, the fibres of a pale brown colour.

CAULIS erectus, rigidus, pedalis, pupureus, ftriatus, foliofus, hirfutus, in quibufdam vix ramofus in aliis ramofiffimus.

STALK upright, rigid, about a foot high, purple, ftriated, leafy, and hirfute, in fome fcarce branched at all, in others very much fo.

FOLIA alterna, feffilia, hirfuta, inferiora obtufe ovata bafi anguftiora, fuperiora angufta, reflexa, tortuofa, ramorum linearia, fuberecta.

LEAVES alternate, feffile, hirfute, the bottom ones of a blunt oval fhape, and narrow at bottom, the upper ones narrow, turning back and twifted, thofe of the branches linear and nearly upright.

FLORES erecti, nunquam fefe explicantes ficut plerique flores Claffis Syngenefiæ, externi purpurei, interni flavefcentes, cum cavitate in medio.

FLOWERS upright, never expanding themfelves like moft of the flowers of the Clafs Syngenefia, externally purple, internally yellow, with a cavity in the middle.

CALYX *communis* imbricatus, fquamis fubulatis, erectis, purpureis, hirfutis, laxis, *fig.* 1.

CALYX: the *common* Calyx compofed of a number of fcales, which are narrow and pointed, upright, purplifh, hirfute, and loofely connected *fig.* 1.

COROLLA *compofita*, radiata; *Corollulæ hermaphroditæ* tubulofæ, numerofæ in difco, *fig.* 2. *femineæ* ligulatæ, pauciores in radio, *fig.* 3. *Propria* hermaphroditi infundibuliformis, flava, limbo quinquefido, *fig.* 2 : *Feminæ* ligulata, linearis, erecta, purpurea, hermaphrodita longior, *fig.* 3.

COROLLA *compound* and radiated ; the *hermaphrodite flowers* tubular and numerous in the middle, *fig.* 2. the *female flowers* ligulate, and fewer in the circumference, *fig.* 3 : each *hermaphrodite* flofcule, funnel-fhaped, yellow, with the limb divided into five fegments, *fig.* 2: each *female* flofcule, linear, upright, purple, longer than the hermaphrodite flower, *fig.* 3.

STAMINA *hermaphroditis* : FILAMENTA quinque, capillaria, breviffima : ANTHERÆ in tubum coalitæ.

STAMINA in the *hermaphrodite flowers* ; five FILAMENTS, very fmall and fhort ; the ANTHERÆ united into a tube.

PISTILLUM *Hermaphroditis* : GERMEN coronatum Pappo corolla paulo longior, *fig.* 4. STYLUS filiformis longitudine Pappi *fig.* 5 ; STIGMA bifidum *fig.* 6 : *Femineis* : GERMEN tenue, Pappo longitudine fere Corollæ, *fig.* 7 ; STIGMATA duo, tenuiffima, *fig.* 8.

PISTILLUM of the *hermaphrodite flowers* ; the GERMEN crowned with a Pappus or Down a little longer than the Corolla, *fig.* 4 ; the STYLE filiform, the length of the Pappus, *fig.* 5 ; STIGMA bifid, *fig.* 6 : of the *Female flowers* ; the GERMEN flender, the Pappus nearly the length of the Corolla, *fig.* 7 ; two STIGMATA very flender, *fig.* 8.

SEMINA oblonga, pallide fufca, *hirfuta, lente auct :* *fig.* 9 : PAPPUS feffilis, lutefcens, fimplex, *fig.* 10.

SEEDS oblong, of a pale brown colour, *hirfute, magnified, fig.* 9 ; PAPPUS feffile, yellowifh and fimple *fig.* 10.

The Erigeron Acre is by no means a common plant in our neighbourhood, yet occurs very frequently on the hilly and chalky ground about *Charlton Wood,* particularly in the chalk pits on the left hand fide of the lane behind the Church.

It flowers in the months of Auguft and September, and is confidered as a pretty fure indication of a barren foil. It has a tafte fomewhat warm and biting, and hence has received its name of *Acris.*

We have rather chofen to retain LINNÆUS's name of *Erigeron* than adopt RAY's name of *Fleabane,* which tends to confound it with the Genus *Conyza.*

It frequently grows much taller, and is often found much fmaller than the fpecimen we have figured.

SENECIO VULGARIS. GROUNDSEL.

SENECIO *Linnæi Gen. Pl.* SYNGENESIA POLYGAMIA SUPERFLUA. *Receptaculum* nudum. *Pappus* fimplex.

Calyx cylindricus, calyculatus ; *fquamis apice fphacelatis.*

Raii Syn. HERBÆ FLORE COMPOSITO, SEMINE PAPPOSO NON LACTESCENTES, FLORE DISCOIDE.

SENECIO *vulgaris* corollis nudis, foliis pinnato-finuatis amplexicaulibus, floribus fparfis. *Linn. Syft. Vegetab.*

p. 630. *Sp. Pl.* 1216. *Fl. Suecic.* p. 290.

SENECIO corollis nudis, foliis pinnato-finuatis amplexicaulibus, floribus fparfis. *Haller hift.* n. 58.

SENECIO *vulgaris. Scopoli Fl. Carniol.* p. 162. n. 1063. *Hudfon Fl. Angl.* p. 315.

SENECIO minor vulgaris. *Baubin Pin.* 181.

SENECIO *vulgaris. Parkinfon* 671.

ERIGERON *Gerard. emac.* 278. *Raii Syn.* p. 178. Common Groundfel or Simfon.

RADIX annua, e plurimis fibrillis albidis conftans.

CAULIS fimplex, erectus, pedalis, ramofus, fæpe purpureus, fubangulofus, in junioribus plantis verfus apicem fubtomentofus.

FOLIA obfcure virentia, glabra, amplexicaulia, pinnato-finuata, pinnis acute dentatis.

PEDUNCULI ftriati, uniflori, primum erecti, peracta florefcentiâ penduli, demum erecti.

CALYX : communis primum cylindraceus, demum conicus ; *Squamis* fubulatis, plurimis, in cylindrum fuperne contractis parallelis, contiguis, æqualibus, paucioribus bafin imbricatim tegentibus, apicibus omnium nigricantibus, *fig.* 1.

COROLLA *Compofita,* longitudine calycis ; *Corollulæ* hermaphroditæ, tubulofæ, numerofæ in difco, infundibuliformes ; *limbo* reflexo, quinquefido : Radio nullo, *fig.* 2, 3.

STAMINA : FILAMENTA quinque, capillaria, minima ; ANTHERA cylindracea, tubulofa.

PISTILLUM : GERMEN ovatum ; STYLUS filiformis, longitudine ftaminum ; STIGMATA duo oblonga, revoluta.

SEMEN oblongum, ftriatum, fufcum ; PAPPUS fimplex, albus, femine triplo fere longior, *fig.* 4 ; RECEPTACULUM nudum, fcabrum.

ROOT annual, confifting of numerous white fibres.

STALK fingle, upright, about a foot high, branched, often purple, flightly angular, in the young plants, towards the top, thinly covered with down.

LEAVES of a deep and dull green colour, fmooth, embracing the ftalk, pinnato-finuated, the pinnæ fharply indented.

PEDUNCLES ftriated, fupporting one flower on each, at firft upright, when the flowering is over they become pendulous, and laftly upright.

CALYX ; the common Calyx firft cylindrical and laftly conical ; the *Squamæ* fubulate, numerous, contracted above into a Cylinder, parallel, contiguous and equal ; thofe at the bafe of the calyx fewer, lying one over another, the tips of all of them blackifh, *fig.* 1.

COROLLA *Compound,* the length of the Calyx ; the *Florets* hermaphrodite, tubular and numerous in the difk or middle, funnel-fhaped, the *limb* reflex and divided into five fegments : the Radius wanting, *fig.* 2, 3.

STAMINA : FILAMENTS five, capillary, and very minute : ANTHERÆ united into a tube.

PISTILLUM : GERMEN oval ; STYLE filiform the length of the Stamina ; STIGMATA two, oblong, and bent back.

SEED oblong, ftriated and brown ; the PAPPUS fimple, white, almoft three times the length of the feed, *fig.* 4 ; RECEPTACLE naked, and rough.

The Groundfel is a Plant which is well known to grow exceedingly common in Gardens, cultivated Ground, and on Walls, flowering all the year if the weather be mild.

Although it is fcarcely ufed at prefent as a medicine, yet according to fome Authors it is not without confiderable virtues : the juice, or decoction of it taken internally, operates gently by vomit ; and the plant externally applied, is faid to be ufeful in inflamed Breafts, the Scrophula, and other Inflamations.

Mr. RAY fufpects that it might be given with advantage in Worms, as Farriers and Horfe-dealers give the juice of it to Horfes that are troubled with thofe kind of Worms called *Bottes,* and to which it is prefently fatal.

Birds of various kinds are fond of the feeds and tops of this plant ; and a great variety of *Caterpillars* particularly thofe of the *Phalæna Jacobeæ* eat it readily.

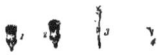

Senecio vulgaris

BELLIS PERENNIS. COMMON DAISY.

BELLIS. *Linnæi Gen. Pl.* SYNGENESIA POLYGAMIA SUPERFLUA.
      Raii Synopsis Gen. S. HERBÆ FLORE COMPOSITO DISCOIDE, SEMINIBUS PAPPO DESTITUTIS,
      CORYMBIFERÆ DICTÆ.

BELLIS   *perennis,* scapo nudo. *Linnæi System. Vegetab* p 640. *Fl. Suecic. p.* 296. *Haller hist. p.* 39. *Scopoli.*
      Fl. Carniol. v. 2. 146.

BELLIS   sylvestris minor *Bauhin pin.* 261. *Gerard emac.* 635. *Parkinson* 530. *Raii Syn. p.* 184. *Hudson Fl.*
      Angl. 320. *OEder. Fl. Dan. icon.* 503.

RADIX perennis, fibrosa.	ROOT perennial, and fibrous.
FOLIA ovata, dentata, hirsutula, in petiolos longos decurrentia ; disrupta fila trahentia.	LEAVES oval, indented, slightly hirsute, running down the footstalks, which are long and if broke across appear stringy.
SCAPI teretes, hirsuti, triunciales, uniflori, ad apicem fistulosi.	STALKS round, hirsute, about three inches high, supporting one flower, at top hollow.
CALYX *communis* simplex, foliolis æqualibus *fig.* 1. apice membranaceis, hirsutis, obtusis *fig.* 2. lente auct.	CALYX the *common* calyx simple, the leaves equal, *fig.* 1. at the top membranous, hairy and obtuse, *fig.* 2. one of the tips magnified.
COROLLA *composita,* radiata: *Corollulæ* hermaphroditæ, tubulosæ, numerosæ in disco. *Fœmininæ* ligulatæ, calycis foliis plures in radio. Flosculi *Hermaphroditi* infundibuliformes quinquefidi flavi, *fig.* 3, 4. lente auct : *Fœminei* ligulati, lanceolati, albi, *fig.* 10.	COROLLA *Compound* and radiated : the *Corollulæ* or flosculi in the disk or middle numerous, tubular, and *hermaphrodite,* those in the radius or circumference flat, more numerous than the leaves of the calyx, and *female.* the *Hermaphrodite* Flosculi funnel shaped, divided into five segments and yellow, *fig.* 3, 4. magnified. The *Female* flosculi tubular at bottom, flat towards the extremity, lanceolate, and white, *fig.* 13.
STAMINA *Hermaphroditis*: FILAMENTA quinque brevissima, *fig.* 5. ANTHERA cylindracea, tubulosa, *fig.* 6.	STAMINA in the *Hermaphrodite* flower: five FILAMENTS very short, *fig.* 5. ANTHERÆ united into a tube, *fig.* 6.
PISTILLUM *Hermaphroditis*: GERMEN ovatum, *fig.* 9. STYLUS filiformis, *fig.* 8. STIGMA crassiusculum, bifidum, *fig.* 7. *Fœmineæ*: GERMEN ovatum, *fig.* 12. STYLUS filiformis. STIGMATA duo patula, linearia, *fig.* 11.	PISTILLUM of the *Hermaphrodite* flower: GERMEN oval, *fig.* 9. STYLE thread-shaped, *fig.* 8. STIGMA thickish and bifid, *fig.* 7. of the *Female* flower: GERMEN oval, *fig.* 9. STYLE thread-shaped, two STIGMATA narrow and spreading, *fig.* 11.
SEMINA ovata, compressa, marginata, pappo destituta, *fig.* 14.	SEEDS oval, flat, margin'd without any pappus or down, *fig.* 14.
RECEPTACULUM nudum, conicum, *fig.* 15.	RECEPTACLE naked and of a conical figure, *fig.* 15.

The Daisy has been recommended by some writers to be given in hectic fevers, caused by drinking cold water when the blood has been heated by exercise, either infused in water or milk.

In some parts of Germany, it is said to be boiled and eaten with meat as a pot-herb; but it does not seem to promise much either as physic or food for man. Sheep and horses refuse it, and it is very probable that none of our cattle eat it willingly; if so the owners of lands pay dear for their enamelled meads, and daisied carpets, but this part of husbandry seems as yet little understood or attended to. As rural œconomists we have ventured to say thus much in dispraise of this flower, notwithstanding the lavish encomiums the father of our English poets has bestowed on it :

    —————— In special one called Se of the daie
    The Daisie, a floure white and rede,
    And in french called *La bel Margarete*
    O commendable floure, &c.
    —————— Above all flouris in the mede
    Than love I most those flouris white and rede ·
    Such that men call in Daisies in our Town.

Chaucer is perhaps the first that takes notice of the Horologium Floræ or opening and shutting of flowers at a particular time of the day.

    —————— She that is of all flouris the floure,
    Fullfilled of all virtue and honoure ;
    And ever alike fair and fresh of hewe,
    As well in winter as in summer newe,
    As soon as ever the Sunne ginneth West
    To sene this floure, how it will go to rest,
    For fear of night so hateth she darknesse
    Her chere is plainly spread in the brightnesse
    of the Sunne. ——————
    Well by reason men it calle maie
    The Daisie, or else the Eye of the Daie
    And at the last there tho began anon
    A Lady for to sing right womanly
    A Bargonet in praising the Daisie
    For as methought among her notis swete
    She said *Si douce est la Margarete*

Retuned by Dryden in his own numbers :

    And then the Band of Flutes began to play,
    To which a Lady sung a Virelay ;
    And still at every close she would repeat
    The Burden of the Song the *Daisy is so sweet*
    The Daisy is so sweet when she begun
    The troops of Knights and Dames continued on
    The Consort and the voice so charm'd my Ear
    And sooth'd my Soul that it was Heaven to hear.

Etymologists agree with the Old Bard in his derivation of the Daisy, viz. Days Eye. Under the French name Margarette it is probable a compliment was intended to some lady, but Critics are not agreed who this lady was. Like many other flowers the Daisy becomes double by culture, and frequently *proliferous,* in this state it is called the *Hen and Chicken* Daisy.

Viola odorata.

VIOLA ODORATA. SWEET VIOLET.

VIOLA *Linnæi Gen. Pl.* SYNGENESIA MONOGAMIA.

Calyx pentaphyllus. *Corolla* pentapetala, irregularis, poſtice cotnuta; *Capſula* ſupera, trivalvis, unilocularis.

Raii Syn. Gen. 24. HERBÆ PENTAPETALÆ VASCULIFERÆ.

VIOLA *odorata* acaulis, foliis cordatis, ſtolonibus reptantibus, bractæis ſupra medium pedunculi.

VIOLA *odorata*, acaulis, foliis cordatis, ſtolonibus reptantibus. *Linn. Syſt. Vegetab. ſ:* 668.

VIOLA acaulis ſtolonifera, foliis cordatis, *Haller hiſt. helv. n.* 558

VIOLA *odorata, Scopoli Fl. Carniol: n.* 1097.

VIOLA martia purpurea flore ſimplici odoro. *Bauhin Pin: p:* 199. martia alba. *p.* 199.

VIOLA nigra ſive purpurea. *Ger: emac.* 550:

VIOLA ſimplex martia. *Parkinſon* 755: *Raii Syn: p.* 364: Purple Sweet Violet, and White Sweet-ſcented Violet. *Oeder Fl. Dan: icon.* 309.

RADIX perennis, fibroſa, albida, in ſeneſcente plantâ baſi petiolorum quotannis reliĉtâ pars ſuperior radicis tuberculoſa evadit, et ſupra terram emi-net; e ſinu horum nodorum naſcuntur ſtolones, qui humi repent, et foliis inſtruuntur ſtipulis-que ejuſdem formæ ac illæ quæ ad baſin plantæ inveniuntur.

FOLIA ſubrotundo-cordata, crenata, ſuperne glabra, in-ferne hirſutula, junioribus involutis.

STIPULÆ radicales, ovato-lanceolatæ, membranaceæ, ſerratæ, dentibus glanduliferis.

PEDUNCULI radicales, infra Bracteas quadrangulares, ſupra Bracteas dorſo canaliculati, apice incur-vati, uniflori.

BRACTEÆ duæ, lanceolatæ, plerumque oppoſitæ, ap-preſſæ, ſupra medium pedunculi.

CALYX: PERIANTHIUM pentaphyllum, perſiſtens, fo-liolis oblongo-ovatis, obtuſis, e viridi purpuraſ-centibus, *fig.* 1.

COROLLA pentapetala, irregularis, violacea, odorata, petalum infimum Nectario corniculato, obtuſi-uſculo, apice compreſſo inſtructum, Petala la-teralia prope baſin barbata, *fig.* 2.

STAMINA: FILAMENTA quinque breviſſima ægre diſ-tinguenda: ANTHERÆ flaveſcentes, biloculares, vix connexæ, membranâ ovato acuta aurantiaca terminatæ; e parte poſteriori duarum Anthera-rum exit Nectariumque intrat appendicula viridi, linearis, compreſſa, *fig.* 5. 4. 3.

PISTILLUM: GERMEN ſubrotundum; STYLUS baſi tenuior et paululum tortuoſus; STIGMA unci-natum, Antheris paulo longius, *fig.* 6, 7.

PERICARPIUM priuſquam dehiſcit, ſubrotundo-tri-angulare, purpuraſcens, villoſum; trivalve valvulis ſubrotundis concavis, *fig.* 8.

SEMINA plurima, rotunda, nitida, ſtraminea, appendi-culata, *fig.* 9.

ROOT perennial, fibrous and whitiſh; in old plants the upper part of the root becomes knobby, and appears above ground, the knots or knobs being formed from the bottoms of the foot-ſtalks of the leaves which are yearly left; from the boſoms of theſe knobs ſpring the ſtolones or ſhoots which creep on the ground, and are furniſhed with leaves and the ſame kind of Stipulæ which are obſervable at the bottom of the plant.

LEAVES heart-ſhaped and ſomewhat round at the tip, crenated, on the upper ſide ſmooth and ſhining, underneath ſlightly hairy, when young rolled in at the edges.

STIPULÆ ſpringing from the root, ovato-lanceolate, membranous, ſerrated at the edges, each ſerra-ture or tooth terminating in a minute gland.

PEDUNCLES ſpringing from the root, below the Bracteæ quadrangular, above the Bracteæ grooved on the upper ſide, at top incurvated, ſupporting one flower.

BRACTEÆ two, lanceolate, generally oppoſite to each other, preſſed to the ſtalk, and *placed above the middle of the Peduncle.*

CALYX: a PERIANTHIUM of five leaves, continuing, each leaf of an oblong oval ſhape, obtuſe at the tip, and of a greeniſh purple colour, *fig.* 1.

COROLLA: of five PETALS, irregular, of a bluſh pur-ple colour and ſweet ſmell, the lowermoſt ter-minating in a blunt horned NECTARIUM, a little flattened at the extremity, the two ſide Petals bearded near the baſe, *fig.* 2.

STAMINA: five FILAMENTS ſo ſhort as hardly to be diſtinguiſhed; ANTHERÆ yellowiſh, bilocular, ſcarcely connected together, terminated by an oval-pointed, orange-coloured membrane; from the back of two of the Antheræ, ſprings a ſlend-er, flat, greeniſh appendage, which enters the Nectarium, *fig.* 5, 4, 3.

PISTILLUM: GERMEN rouudiſh; STYLE ſlendereſt at bottom and a little twiſted; STIGMA hooked, and a little longer than the Antheræ, *fig.* 6, 7.

SEED-VESSEL, before it burſts, roundiſh, rather ap-proaching to triangular, of a purpliſh colour, and villous appearance, ſplitting into three round-iſh hollow valves, *fig.* 8.

SEEDS ſeveral, round, ſhining, of a ſtraw colour, ter-minated by a little appendage, *fig.* 9.

The *Viola odorata* delights to grow under warm hedges, particularly near Woods; if the Spring be favourable, it is generally in full bloom in the month of March, and towards the latter end of Summer ripens its ſeeds. A variety of this plant frequently occurs with a white flower, not inferior in its agreeable ſcent to the blue one; and ſometimes this plant is found double, in which ſtate it is often introduced into Gardens, and being furniſhed with abundance of creeping ſhoots, it is by means of theſe propagated with the utmoſt facility.

This ſpecies of Violet bears a conſiderable reſemblance to the *Viola hirta*, the mode of diſtinguiſhing them we ſhall point out when we deſcribe the latter.

A ſyrup made from the flowers is uſually kept in the ſhop, and frequently given to children where a gentle laxative is required: it is likewiſe in uſe as a teſt to try acid and alcaline ſubſtances.   The

The feeds are faid by Authors to poffefs a diuretic quality, and hence the powder of them has been recommended in the ftone and gravel.

The great Bacon, who frequently defcended from his fublimer ftudies, and amufed himfelf with enquirie into the qualities and properties of plants, has left us a curious method of preferving the fcent of this flowe:

"*Take Violets, and infufe a good pugil in a quart of Vineger, let them ftand three quarters of an hour, and tal*
"*them forth, and refresh the infufion with like quantity of Violets feven times; and it will make a Vineger fo fre*
"*of the flower, as if a twelve moneth after it be brought you in a faucer, you fhall fmell it before it come at yor*
"*Note. It fmelleth more perfectly of the flower a good while after than at the first.*"

The illuftrious prefcriber has given no directions concerning the ufe of this preparation, but it appears to 1 ro be one of the moft grateful prefervatives againft infection, efpecially if the ftrongeft diftilled vinegar whic has been drawn over in glafs, be made ufe of.

The Violet has been much complimented by the antient Poets; and our Shakespeare gives it a confp ⟨c⟩uous place in his catalogue of flowers.

———————————————— "*Violets dim,*
"*But fweeter than the lids of Juno's eyes,*
"*Or Cytherea's breath.*"

The Commentators have not been fuccefsful in informing us how the "*lids of Juno's eyes*" bear any refembl⟨ant⟩ to "*Violets dim,*" not recollecting that ιοβλεφαρος *(having violet eyelids)* was a complimentary title with the Greek poet This epithet alludes to a well known cuftom which ftill prevails in Greece, of colouring the eye lids blue. *"
"Grecian girl is painted blue round the eyes; and the infides of the fockets, with the edges on which th
"lafhes grow, are tinged with black: For colouring the lafhes and focket of the eye, they throw incenfe ⟨⟩
"Gum of Labdanum on fome coals of fire, intercept the fmoak which afcends, with a plate, and collect th
"foot: This I faw applied; a girl fitting crofs-legged, as ufual, on a fopha, and clofing one of her eyes, too
"the two lafhes between the fore finger and thumb of her left hand, pulling them forward, and then thruftin
"in, at the external corner, a bodkin which has been immerfed in the foot, and extracting it again, the pa
"ticles before adhering to it remained within, and were prefently ranged round the organ, ferving as a foil 1
"its luftre, befides contributing, as they fay, to its health, and increafing its apparent magnitude," Chandler *Travels into Greece.*

Altho' the poet of nature has been rather obfcure on this fubject, where he copies the ancients; he mak⟨e⟩ ample amends when he gives us the genuine effufions of his own imagination. With what precifion ar delicacy does he defcribe the foft enchantment of plaintive mufic, as refembling the fweetnefs of this flower illuftrating in a beautiful fimile the object of one fenfe by that of an other!

"*That ftrain again;——it had a dying fall;*
"*Oh! it came o'er my ear, like the fweet fouth,*
"*That breathes upon a* bank *of violets,*
"*Stealing and giving odour!*"

*A Greek poet fuppofed to be a Chriftian, from the feverity of his manners and purity of his inftructions, forbids this cuftom of painting the ey lids, in the rules of conduct which he addreffes to young women,

" Μεδε μιλαιμι ταιτω τας βαρδαχεως οπωπας."

Naumachius.

It is probable that the Greeks borrowed this fafhion from their Afiatic neighbours; Jerenel, a native of Zidon, put her eyes in painting, as t tranflators tell us in the margin of our bible; the Prophets alfo allude to and cenfure this cuftom, fee *Jeremiah* iv. 30. *Ezechel* xxiii 40.

Viola hirta.

VIOLA HIRTA. HAIRY VIOLET.

VIOLA *Linnæi Gen. Pl.* SYNGENESIA MONOGAMIA.

Calyx pentaphyllus. *Corolla* pentapetala, irregularis, poſtice cornuta. *Capſula* ſupra, trivalvis, unilocularis.

Raii Synop. Gen. 24. HERBÆ PENTAPETALÆ VASCULIFERÆ.

VIOLA *hirta* acaulis, foliis petioliſque hirſutis, bracteis infra medium pedunculi.

VIOLA *hirta* acaulis, foliis cordatis piloſo hiſpidis. *Linn. Syſt. Vegetab. p.* 668.

VIOLA acaulis, foliis cordatis hiſpidis. *Haller hiſt. helv. n.* 559.

VIOLA *hirta* *Hudſon Fl. Angl. p.* 330.

VIOLA martia major hirſuta inodora. *Hiſt. ox. II.* 475.

VIOLA trachelii folio vulgo. *Raii hiſt.* 1051. *Syn. p.* 365. Violet with Throat-wort leaves.

So great is the ſimilarity betwixt this Species and the *Viola odorata*, that to deſcribe it in the ſame manner as I have that plant, would be to repeat nearly the ſame words. To avoid this ſameneſs of expreſſion, I ſhall adopt a deſcription in the way of contraſt, which will enable me to point out the differences of each in a manner more ſtriking, and I hope equally ſatisfactory to my botanic readers.

I would firſt premiſe, that as it is my greateſt wiſh to clear up every difficulty reſpecting the ſpecies and varieties of thoſe plants which come properly before me, ſo I have with that view, not only examined this plant with the greateſt attention, where it has grown wild, but alſo cultivated it in my garden along with the *odorata*, and hence, ſeeing and noticing its mode of growth throughout the year, have perhaps been able to obtain a clearer idea of its hiſtory, than thoſe who may have viewed it at one particular ſeaſon only.

The *Viola odorata* throws out from the upper part of its root a number of ſtolones or ſhoots, which trail on the ground, and quickly take root at the joints, whereby it propagates itſelf very faſt: the *hirta* likewiſe encreaſes itſelf by throwing out young ſtalks; but then they are not procumbent, nor do they ever ſtrike root as thoſe of the *odorata* do; hence the *hirta* does not encreaſe ſo faſt, nor ſpread ſo wide. Although LINNÆUS makes a conſiderable difference in the form of the roots of theſe plants, yet from what I have obſerved, this difference proceeds chiefly from the age of the roots; for in both ſpecies, the older they are, the more full are they of tubercles or cicatrices, formed by the annual ſhedding of the leaves.

The radical *Stipulæ* are lanceolate and ſerrated in both ſpecies.

The *footſtalks* of the leaves form perhaps the moſt obvious difference; in the *odorata* they are nearly ſmooth; in the *hirta* they are very hirſute, and this hairineſs puts on a kind of ſilvery appearance in the young plants of this ſpecies, where it is remarkably conſpicuous.

In the leaves themſelves the difference is for the moſt part, not very remarkable, for in both ſpecies they are ſomewhat hirſute underneath; thoſe of the *hirta* however, are ſometimes remarkably ſo, from growing in particular ſoils or ſituations: the leaves of the *odorata* have a more gloſſy appearance on their upper ſurface, but this ſcarce diſcriminates them unleſs they are contraſted. With reſpect to ſhape and ſize likewiſe, the difference is not very obvious; both ſpecies when in bloom are ſmall, compared with the ſize to which they afterwards grow. In make they are ſomewhat longer, and not ſo perfectly heart-ſhaped.

In the ſpecimens of this plant which I have examined, I could not perceive that ſenſible difference which LINNÆUS notices *(vid. Mantiſſ. Plant. alt. p.* 483.) in the ſhape of the *Peduncle* above the Bracteæ; in both ſpecies they certainly are channeled at the back: in the ſituation of the Bracteæ, however, there is a very conſiderable difference, which does not appear to have been taken notice of, and this ſeemed to me to be ſo obvious a character, that I truſt it will apologize for my altering its ſpecific deſcription: in the *odorata*, the *Bracteæ* are placed above the middle of the Scapus, or Peduncle; in the *hirta*, they are ſituate below it: but there is one caution neceſſary to be obſerved reſpecting this character, viz. that the Bracteæ of each be obſerved, juſt when the flowers are fully expanded, for as that part of the Scapus, which is ſituated above the Bracteæ, grows conſiderably longer by the time that the flowers of the *odorata* are faded, ſo they ſhould both be examined when of an equal age, otherwiſe this diſtinction will not appear ſo remarkable.

The flowers of the *hirta*, in general, appear about a week later than thoſe of the *odorata*, are of a paler blue colour, and entirely want that ſweet fragrance which renders the *odorata* ſo grateful a harbinger of the Spring. In the other parts of the fructification, theſe plants are very ſimilar to each other; but there is one circumſtance reſpecting the manner in which they produce and diſperſe their ſeeds, which may not be generally known.

LINNÆUS in his *Flora Suecica, n.* 789, obſerves that the flowers which the *Viola mirabilis* firſt produces from the root, are furniſhed with Petals, yet that theſe for the moſt part are barren, while thoſe which blow later the ſame Spring, and riſe from the ſtalk, although deſtitute of Petals, produce perfect ſeed: and JACQUIN, in his excellent work the *Flora Auſtriaca*, where this plant is figured, *(Vol.* 1. *tl.* 19.) confirms the truth of LINNÆUS'S obſervations, and ſays that the barreneſs of thoſe flowers appeared to ariſe from a deficiency of the Stylus. LINNÆUS in his valuable treatiſe above quoted, obſerves likewiſe, that the flowers of the *Viola montana*, which appear firſt, are furniſhed with Petals, but that thoſe which are afterwards produced have no Petals, yet nevertheleſs are fertile; and this I find, on repeated examination, to be the caſe with the *Viola odorata* and *hirta*, but more particularly the latter: they differ from the *Viola mirabilis* in this reſpect, that all the flowers which are formed, both with and without Petals, produce perfect ſeed. I was led to this diſcovery from obſerving a ſingle plant of the *Viola hirta*, to produce about the middle of Summer, ten or twelve capſules of ripe ſeeds, on which I was certain in the Spring no more than two or three bloſſoms had appeared: the next Spring I diſcovered, that beſides thoſe perfect bloſſoms which firſt ſpring up, this plant continues for a month or more to throw out new flowers, which are entirely deſtitute of Petals, or have only the rudiments of them which never appear beyond the Calyx; but all the other parts of the fructification are perfect. The capſules in both theſe ſpecies, when they become nearly ripe, lay cloſe to the ground, ſo that when they burſt, the ſeeds have an eaſy acceſs into the earth.

There is ſome difference with reſpect to the ſoil and ſituation in which theſe two plants delight; the *odorata* grows very generally under warm hedges, and in woods; the other appears to be pretty much confined to a chalky ſoil, and often occurs in more expoſed ſituations, in the fields and on the banks about *Clapton*, it may be found in tolerable abundance.

Viola tricolor

VIOLA TRICOLOR. WILD PANSIE.

VIOLA   *Linnæi Gen. Pl.* SYNGENESIA MONOGAMIA.
     Raii Synop. Gen. 20. HERBÆ PENTAPETALÆ VASCULIFERÆ.
VIOLA   tricolor, caule triquetro diffuſo, foliis oblongis inciſis, ſtipulis pinnatifidis. *Linn. Syſt. Vegetab. p.*
     668. *Fl. Suecic.* 307.
VIOLA   caule diffuſo, ramoſo, foliis ovatis dentatis, flore calyce paulo majori. *Haller. hiſt. tom.* 1. *n.* 569.
VIOLA   bicolor arvenſis. *C. Bauhin. pin.* 200.
VIOLA   tricolor ſylveſtris. *Parkinſon.* 755.
JACEA   bicolor frugum et hortorum vitium. *I. Bauhin.* III. 548. *Raii Syn. p.* 336. 11. *Hudſon. Fl.*
     Angl. p. 331. *Scopoli. Fl. Carniol. p.* 183.

RADIX ſimplex, fibroſa.

ROOT ſimple and fibrous.

CAULIS palmaris et ultra, plerumque diffuſus, ramoſus, anguloſus, ad baſin ſordide purpureus ; rami alterni.

STALK about four or ſix inches high, generally ſpreading, branched, angular, at bottom of a dull purple colour ; the branches alternate.

FOLIA longe petiolata, elliptica, crenata, inferioribus ſæpe minoribus, ſubrotundis, ſuperioribus anguſtis, ſubdentatis.

LEAVES placed on long foot-ſtalks, elliptical, crenated, the lowermoſt often ſmaller and roundiſh, the uppermoſt narrow and ſlightly indented.

STIPULÆ ad baſin laciniato-pinnatifidæ, laciniis linearibus, extrema oblonga, dentata.

STIPULÆ at bottom jagged and pinnatifid, the laciniæ or jags linear, that which terminates the Stipula oblong and indented.

PEDUNCULI ſubquadrangulares, alterni, apice incurvati, dorſo canaliculati, ſtipulis duobus parvis, membranaceis, prope florem, inſtructi.

FOOT-STALKS of the flowers, nearly quadrangular, alternate, bent in at top, channeled on the back, and furniſhed with two ſmall membranous Stipulæ near the flower.

CALYX : PERIANTHIUM pentaphyllum, perſiſtens, foliolis acutis, trin *ſuperiora* minora, ad baſin æqualia, ſuprema erecta, petalis ſupremis longiora, duo *inferiora* apice et baſi cæteris longiora, baſique latiora, petalis infimis breviora, *fig.* 2.

CALYX : a PERIANTHIUM of five leaves and continuing, the leaves ſharply pointed, the three *upper* ones ſmalleſt, and equal at bottom, the uppermoſt upright and longer than the uppermoſt petals, the two *under leaves* longer both at bottom and top than the reſt, and at bottom likewiſe broader, ſhorter than the lowermoſt petals, *fig.* 2.

COROLLA pentapetala, irregularis, duo ſuperiora ſubrotunda, integerrima, albida, deorſum ſpectantia ; lateralium *lamina* ovata, obtuſa, ad baſin barbata, lineuque brevi purpurea notata ; infimum latum emarginatum, ad baſin flavum, lineis quinque purpureis pictum, CALCARE SEU NECTARIO

COROLLA pentapetalous and irregular, the two uppermoſt petals roundiſh, entire, and reflected ; the lamina or broad part of the ſide petals oval, obtuſe, bearded at bottom, and marked with a ſhort purple line ; the lowermoſt petal broad, emarginate, yellow at bottom, and ſtreaked with five purple lines, and terminated by a

NECTARIUM. terminatum, longitudine calycis, apice violaceo, obtuſo, *fig.* 3, 4, 5, 6.

NECTARY Spur or NECTARY the length of the Calyx, with a blueiſh and blunt point, *fig.* 3, 4, 5, 6.

STAMINA: FILAMENTA quinque, breviſſima ; ANTHERÆ albidæ, vix coadunatæ, biloculares, membranâ croceâ terminatæ, e dubious inferioribus exeunt, nectariumque intrant, appendiculæ duæ lineares, *fig.* 7, 8, 9, 10.

STAMINA : five FILAMENTS very ſhort ; ANTHERÆ whitiſh, ſcarcely united, bilocular, terminated by a ſaffron coloured membrane ; from the two lowermoſt two linear appendages go off and enter the Nectary, *fig.* 7, 8, 9, 10.

PISTILLUM: GERMEN ſubconicum, *fig.* 11 ; STYLUS ad baſin tortuoſus, ſtaminibus longior, *fig.* 12 ; STIGMA capitatum, oblique perforatum, perſiſtens, *fig.* 13.

PISTILLUM : GERMEN ſomewhat conical, *fig.* 11 ; STYLE twiſted at bottom and longer than the Stamina, *fig.* 12 ; STIGMA forming a little head, obliquely perforated and continuing, *fig.* 13.

PERICARPIUM: CAPSULA ovata, glabra, unilocularis, trivalvis, *fig.* 14, 15.

SEED-VESSEL : an oval ſmooth CAPSULE of one cavity and three valves, *fig.* 14, 15.

SEMINA plurima, ovata, fuſca, nitida, appendiculata, valvis ſeriatim affixa, *fig.* 15.

SEEDS numerous, oval, brown and ſhining, with a button to each, affixed in rows to the inſide of the valves, *fig.* 15.

Few plants have acquired a greater variety of names than the *Viola Tricolor* ; in different Authors and different counties we find the following, viz. *Wild Panſie, Herb Trinity, Hearts eaſe, Three faces under a hood, Cull me to you, Love in Idleneſs, &c.* what has occaſioned ſome of theſe is the different appearance it puts on from cultivation and change of ſoil ; in a garden there are few flowers that can boaſt a greater variety or richneſs of colour, few that continue longer in bloſſom, or are cultivated with more eaſe ; it is probable that the large yellow Violet, *Viola lutea,* is no more than a variety of this ſpecies.

The Panſie in its wild ſtate occurs very frequently in cultivated fields, and bloſſoms through moſt of the ſummer months. It is ſo hardy as to appear in Lapland amongſt the few other plants which ornament the waſtes of that Country during its ſhort ſummer. It is eaten by Kine and Goats.

The difference in the form of the Stigma ſeems to divide the plants of this Genus into two families, viz. *Panſies* and *Violets,* in the former the Stigma is round, with a remarkable hole on one ſide of it, in the latter it is hooked.

Linnæus remarks the black lines which ſometimes appear on the Petals, MILTON had obſerved the ſame, " *Panſies freakt with Jet*" In a poor ſoil the purple and yellow in the bloom of this flower frequently become very faint, and ſometimes fade into a perfect white, this variation in colour gives a propriety to the Metamorphoſis of this flower in which SHAKESPEAR pays an elegant compliment to his royal miſtreſs.

> *That very time I ſaw, (but thou could'ſt not!)*
> *Flying between the cold Moon and the Earth,*
> *Cupid all-arm'd : a certain aim he took,*
> *At a fair Veſtal, throned by the weſt,*
> *And looſ'd his love-ſhaft ſmartly from his bow,*
> *As it ſhould pierce a hundred thouſand hearts :*
> *But I might ſee young Cupid's fiery ſhaft*
> *Quench'd in the chaſte beams of the watery moon,*
> *And the imperial votreſs paſſed on,*
> *In maiden meditation fancy-free.*
> *Yet mark'd I where the bolt of Cupid fell,*
> *It fell upon a little weſtern flower ;*
> *Before, milk-white ; now purple with Love's wound,*
> *And Maidens call it Love in Idleneſs.*

OPHRYS APIFERA. BEE ORCHIS.

OPHRYS. *Lin. Gen. Plant. ed.* 3. GYNANDRIA DIANDRIA.

ORCHIS. *Raii Synopſis*, ed. 3. 379. HERBÆ BULBOSIS AFFINES.

OPHRYS. bulbis ſubrotundis, ſcapo folioſo, nectarii labio quinquelobo ; lobis ſubtus inflexis. *Hudſon. Flor. Angl.* 340.

ORCHIS. radicibus ſubrotundis, labello holoſericeo, emarginato, appendiculato. *Haller. Hiſt. Vol.* 2. 1266. *tab.* 24.
    Duas ſpecies *apiferam* et *muſciferam* HUDSONIS ET HALLERI ſub uno nomine *Inſectiferæ* conjungit
    CL. LINNÆUS.

Fuſchii icon 560. *Bauhin pin.* 83. *Gerard, emac.* 212.

RADIX. Bulbi duo, ſubrotundi, inæquales, radiculis longis vix fibroſis ſupra inſtructi.

ROOT. Two roundiſh, inequal bulbs, furniſhed at top with a few ſmall longiſh fibres, but little branched.

CAULIS Semipedalis aut pedalis, teres, *fig.* 1. folioſus.

STALK from half a foot to a foot high, round, *fig.* 1, leafy.

FOLIA Vaginantia, ovato lanceolata, ſubtus ſubargentea, fibris lineata, ſæpe mutilata et fuſca.

LEAVES embracing the ſtalk, of an oval pointed ſhape, underneath ſilvery, with linear fibres, frequently imperfect, and of a brown colour.

BRACTEÆ *magnæ, vaginantes,* virides, longitudine floris.

FLORAL LEAVES *large, in the form of a ſheath,* green, and of equal length with the flowers.

FLORES a tribus ad ſex, ſpicanti.

FLOWERS from three to ſix growing in a ſpike.

COROLLA. PETALA quinque, tria exteriora reliquis majora, ovata, concava, reflexa, purpuraſcentia, ſerioribus pallidioribus, ſubcarinata, carina viridi, *fig.* 2; duo interiora exterioribus quadruplo minora, anguſta, *hirſuta, poſtice canaliculata, ad baſin latiora,* antrorſum extantia.

COROLLA. five PETALS, the three exterior larger than the reſt, oval, concave, turning back, purpliſh, ſomewhat keel ſhaped, the keel green, *fig.* 2. the latter flowering paleſt ; the two interior four times ſmaller than the others, narrow, *hairy, hollow behind, broadeſt at bottom,* and projecting forward.

NECTARII Labellum amplum, leniter convexum, ſuborbiculatum, fuſco ſericeo, maculis flavis frequenter variegatum, quinquelobum, *lobis inflexis, fig.* 3. *lateralibus ſubtriangularibus, hirſutis; fig.* 4. medio anteriorum productiore, apice recurvato flavo, *fig.* 5.

    Machina ſtaminum ſive Stylus longa, ſuberecta, *apice incurvata et ſurſum recurvata, fig.* 11. antice bilocularis, loculis apertis, *fig.* 12, anguſtis, marginibus albis, membranaceis. *fig.* 13.

NECTARY. The lip of the Nectary, large, ſomewhat convex, roundiſh, of a ſilky brown colour, frequently variegated with yellow ſpots ; having five lobes, *the lobes bending underneath,* fig. 3. *The two ſide lobes, ſomewhat triangular and hairy,* fig. 4. the middle of the anterior running out to a point, which turns back, and is of a yellow colour, *fig.* 5. the Style which in this plant ſupports the Stamina, long, upright, *at the tip bending downwards and again upwards,* fig. 11. anteriorly having two cavities which are open and narrow, *fig.* 11. the edges white and membranous, *fig.* 13.

STAMINA. FILAMENTA duo, *fig.* 6. e ſquamula nectariferâ ad baſin Styli exeuntia, nutantia, Stigmati frequenter adhærentia, *fig.* 8. baſi glandula ſive globulo albo pellucido inſtructa, *fig.* 7. ANTHERÆ ſubrotundæ flavæ, *fig.* 9.

STAMINA. two FILAMENTS, *fig.* 6, ariſing from the bottom of the ſtyle out a nectariferous ſcale, hanging down, frequently adhering to the Stigma, *fig.* 8. furniſhed at bottom with a ſmall tranſparent gland or globule, *fig.* 7.—The ANTHERÆ roundiſh and yellow, *fig.* 6.

PISTILLUM. GERMEN oblongum, hexangulare, angulis obtuſis *rectis,* STIGMA, *fig.* 10. melleo liquore obductum, cui particulæ antherarum frequenter adhærent.

PISTILLUM. the GERMEN oblong, having ſix angles, the angles obtuſe, *not twiſted,* the STIGMA, *fig.* 10. covered with a viſcid ſubſtance like honey, to which ſmall particles of the Antheræ frequently adhere.

PERICARPIUM. Capſula oblonga, fuſca, uncialis, *fig.* 14, unilocularis, *fig.* 16, trivalvis, valvis carinatis, *fig.* 15.

SEED VESSEL. a CAPSULE about an inch in length, oblong, brown, *fig.* 14. of one cavity, *fig.* 16, and three valves, the valves keel ſhaped. *fig.* 15.

SEMINA plurima, minuta, oblonga, utraque extremitate membranacea, pellucida, reticulata, *fig.* 18, lente aucta, interiori parti carinæ longitudinaliter affixa *fig.* 17.

SEEDS numerous, ſmall, oblong ; at each end membranous, tranſparent, and reticulated, *fig.* 18. magnified, affixed lengthwiſe to the inſide of the keel of each valve, *fig.* 17.

Flowers in the Months of JUNE and JULY, the Seed is ripe the latter End of AUGUST.

Grows generally on chalky ground near woods, and ſometimes in meadows ; is become ſo rare about *London*, as ſcarcely to be found with any certainty—Mr. *Alchorne* informs me he has frequently gathered it in the pits behind *Charlton* church, and in the woods near *Chiſſelhurſt* in *Kent.*—But it is often met with in plenty at a greater diſtance from town.

The root appears to poſſeſs the ſame virtues with thoſe of the Orchis from which Salop is made, but being much ſmaller, is not worth cultivating on that account. The great reſemblance which the flower bears to a Bee, makes it much ſought after by Floriſts, whoſe curioſity indeed, often prompts them to exceed the bounds of moderation, rooting up all they find, without leaving a ſingle ſpecimen to chear the heart of the Student in his botanic excurſions. The beſt time of tranſplanting them is when they are in flower. This, with moſt of the other Orchis's, was cultivated with the greateſt ſucceſs by the late PETER COLLINSON, Eſq; (whoſe memory will always be revered by every Botaniſt) in his garden at *Mill-hill.*—His method was to place them in a ſoil and ſituation as natural to them as poſſible, and to ſuffer the graſs and herbage to grow round them.

I have not yet heard of their being propagated by ſeed ; it is to be wiſhed that ſome intelligent Gardiner would exert himſelf in making ſome experiments to raiſe them in this way.

Botaniſts have often been at a loſs in claſſing many plants, to find ſome reſemblance by which they might diſtinguiſh their particular ſpecies ; but in this plant the caſe is otherwiſe, the flower is ſo like the inſect that gives it its name, that it ſtrikes every beholder with admiration ; what uſeful purpoſe is intended by it, we do not at preſent know : Some future Obſerver may perhaps diſcover, for they who will examine Nature herſelf, " have much to ſee."

Asplenium Scolopendrium.

ASPLENIUM SCOLOPENDRIUM. HARTSTONGUE.

ASPLENIUM *Lin. Gen.* CRYPTOGAMIA FILICES.

Raii Synopfis. HERBÆ CAPILLARES ET AFFINES.

ASPLENIUM frondibus fimplicibus cordato-lingulatis integerrimis, ftipitibus hirfutis. *Lin. Sp. Pl.* 1537.

ASPLENIUM Frondes lanceolatæ, acuminatæ, bafi cordatæ, integerrimæ, medio latiores. *Scopoli Fl. Carn.*

ASPLENIUM petiolis hirfutis, folio longe lineari-lanceolato, integerrimo, circa petiolum exfciffo. *Haller, hift. n.* 1695.

HEMIONITIS Fufchii Icon. 294

PHYLLITIS vulgaris Cluf. hift.

SCOLOPENDRIA vulgaris Tragi.

LINGUA CERVINA officinarum *Bauhin. pin.* 350. *Gerard. emac.* 1138. *Parkinfon.* 1046. *Raii Synop.* 116. *Hudfon. Fl. Angl.* 384.

RADIX perennis, fibrofiffima, fufca, fibris fibrillis tenuiffimis inftructis.

STIPITES plures, pilofi.

FRONDES cordato-lingulatæ, longitudine pedales, latitudine fere bipollicares, glaberrimæ, margine undulato, nervo medio inferne pilofo.

FRUCTIFICATIO. Glomera linearia, obliqua, in pagina inferiore frondis nervo medio utrinque feriatim difpofita, *fig.* 1. 2. 3.

INVOLUCRUM. Squama linearis, bivalvis, longitudinaliter dehifcens, *fig.* 2.

CAPSULÆ numerofæ, fubglobofæ, uniloculares, pedicellatæ, annulo elaftico cinctæ, *fig.* 5. 7. lente auctæ.

SEMINA numerofa, fubrotunda, minutiffima, *fig.* 7. lente valde auctæ, *fig.* 8.

ROOT perennial, exceedingly fibrous; the fibres brown, and furnifhed with other fibres, which are very minute.

STALKS numerous and moffy, or hairy.

LEAVES tongue-fhaped, at bottom cordate, about a foot in length, and one inch and a half in breadth, of a bright yellowifh green colour and fhining, the margin a little waved, and the mid-rib on the under fide moffy.

FRUCTIFICATION placed in oblique lines on the under fide of the leaf, on each fide of the mid-rib, *fig.* 1. 2. 3.

INVOLUCRUM a linear membrane or cafe, of two valves, opening longitudinally, *fig.* 2.

CAPSULES numerous, ftanding on foot-ftalks, nearly globular, furrounded by an elaftic ring, and having one cavity, as they appear magnified, *fig.* 5. 7.

SEEDS roundifh, very numerous and minute, *fig.* 7. as they appear through a greater magnifier, *fig.* 8.

THIS is one of thofe plants which fome Botanic Writers have called *Epiphyllofpermæ* from producing their feeds on the back of the leaves; LINNÆUS, includes it in his clafs, *Cryptogamia*, as neither Stamina nor Piftilla have yet been difcovered on it. The firft appearance of fructification that we obferve are fome little bags or cafes of a yellowifh or whitifh green colour placed in rows on the under fide of the leaves; *fig.* 1. on opening of which almoft as foon as they become vifible, we find the capfules or feed-veffels, *fig.* 2. very numerous, ftanding upright, and clofe together; at this time they appear of a green colour, as they approach towards maturity they change this for a deep brown; the cafes then open lengthways in the middle, the two fides, by the protrufion of the capfules are turned quite back, and wholly difappear, *fig.* 3. This membranous fubftance or cafe, may be confidered as fimilar to the *Calyptra* in moffes, or *Calyx* in other plants, and ferves to fecure and defend the tender feed and capfules, which being now become ripe, exhibit a moft ftriking proof of that wifdom which the benevolent Author of Nature manifefts in all the works of his creation.

Each capfule or feed-veffel confifts of three parts, firft the foot-ftalk, *fig.* 4. which fupports and connects them to the leaf; fecondly, the jointed fpring, *fig.* 5. which nearly furrounds the third part, or cavity containing the feeds, *fig.* 6. 7.

The feeds being ripe, the cavity containing them is forced open by the elafticity of the jointed fpring, and the feeds fcattered and thrown to a confiderable diftance; one half of the cavity remains connected to one end of the fpring, and the other half to the other end, *fig.* 7.

Some of the capfules being fooner ripe than others, difcharge their feed fooner, fo that it is a confiderable time before they all become empty. On applying an entire row before the microfcope for the firft time, I was immediately ftruck with the motion that appeared in them, and afterwards found that the warmth of my breath occafioned a great number of the capfules to keep continually difcharging their feeds, fo as almoft to give them the appearance of fomething alive. The clofenefs of the capfules to one another affording me but a confufed idea of their ftructure, I feperated them with the point of a penknife, from their connection to the leaf, and again placed them before the microfcope, which then gave me a very different, and after a little examination, a very clear idea of their ftructure; many appeared with the feeds difcharged, feveral in the act of difcharging them, and fome as yet entire; it frequently happened that while I was intently looking at one which I expected would open, at the inftant of difcharging it would be carried out of my fight by the ftrength and elafticity of the fpring, and it was not 'till after repeated trials that I was able clearly to obferve the manner of their opening. The feeds are very numerous, and fcarcely vifible to the naked eye; when magnified, they appear of a roundifh figure, and full of little projecting points.

Both GREW and SWAMMERDAM have given figures on this fubject, but thofe of SWAMMERDAM are by much the moft natural. As a great deal of the fatisfaction in viewing objects of this kind, depends on the goodnefs of the microfcope, that none of my Readers may be difappointed in the experiments they may make with this entertaining inftrument, I may inform them, that the microfcope I make ufe of is that which is fold in the fhops, by the name of ELLIS's *Aquatic Microfcope*, and which is made for this purpofe with particular care and accuracy by GEORGE ADAMS of Fleet-Street, Mathematical Inftrument Maker to his MAJESTY.

This plant may be found in feed from *September* to *November* in fhady lanes and on walls, and is frequently found growing within-fide of old wells. It is moft with but rarely about town, though cultivated in moft of our botanic gardens. The leaves are fubject from a richnefs of foil to be much divided at their extremities, and very much curled at the edges.

It is an officinal plant, and is recommended by RAY from his own experience, as a good medicine againft convulfive diforders.

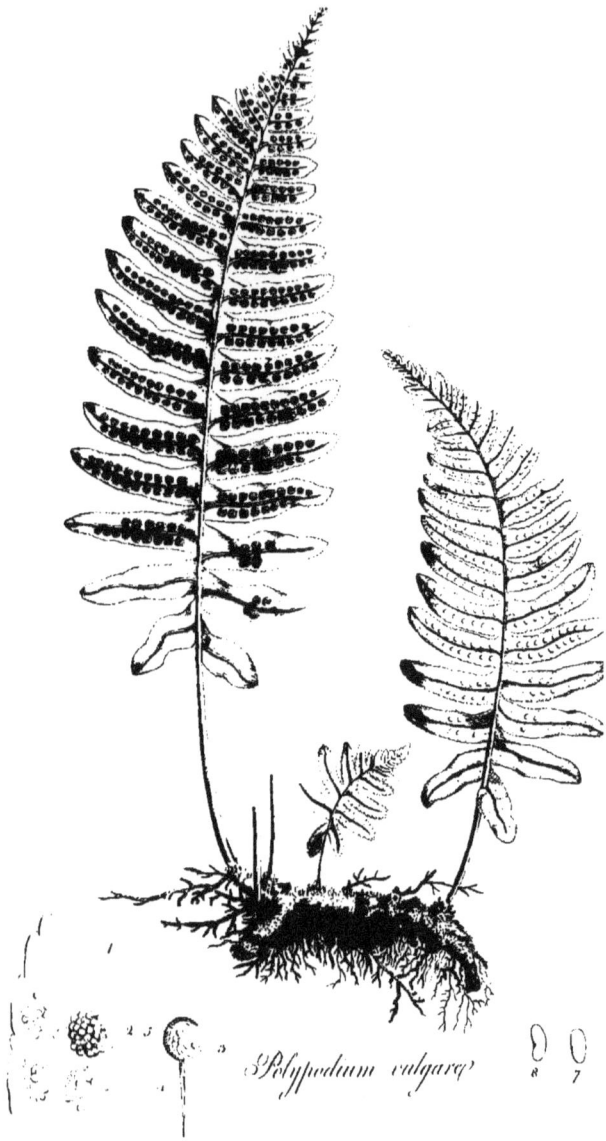

Polypodium vulgare

POLYPODIUM VULGARE. COMMON POLYPODY.

POLYPODIUM *Linnæi Gen. Pl.* CRYPTOGAMIA FILICES.

Fructific. in punctis fubrotundis fparfis per difcum frondis.

Raii Syn. HERBÆ CAPILLARES ET AFFINES.

POLYPODIUM *vulgare* frondibus pinnatifidis : pinnis oblongis fubferratis obtufis. *Linn. Syft. Vegetab.*
. *p.* 786. *Fl. Suecic. p.* 373.

POLYPODIUM foliis pinnatis, lanceolatis, radice fquamata. *Haller hift. n.* 1696.

POLYPODIUM *vulgare. Scopoli Fl. Carniol. n.* 1266.

POLYPODIUM *vulgare. Bauhin. pin.* 359.

POLYPODIUM *vulgare. Parkinfon* 1039.

POLYPODIUM *Gerard emac.* 1138. *Raii Syn. p.* 117, Polypody. *Hudfon Fl. Angl. p.* 387.

RADIX oblique fub terræ fperficie reptat, fibras fuas ex tuberculis quibus plurimis fcatet demittens, ad craffitudinem fere minimi digiti accedens, fquamis fufcis tecta, colore foris buxea, intus fere herbacea, fapore dulci, tandem acerbo et adftringente.

ROOT creeps obliquely under the furface of the earth, fending forth a number of fibres from little tubercles, which are plentifully diftributed over its furface, about the thicknefs of the little finger, fometimes flenderer, covered with brown moffy fcales, externally of a pale yellow colour, internally greenifh, of a tafte at firft fweet, but finally fowerifh and aftringent.

STIPITES læves, interne fulcati.

STALKS fmooth, grooved on the inner fide.

FRONDES femipedales aut pedales, pinnatifidæ, pinnæ oblongæ, fubferratæ, obtufæ, inferne pallidiores.

LEAVES from half a foot to a foot in length, pinnatifid ; the pinnæ oblong, flightly ferrated, obtufe, paleifh underneath.

CAPSULÆ in acervulis, magnis, flavis, rotundis, nervo utrinque feriatim locatæ, pedicellatæ, fubrotundæ, fuperficie granulata a feminibus protuberantibus, annulo elaftico brevi inftructæ, in valvulas duas dehifcentes, *fig.* 2, 3, 3, 5, 6.

CAPSULES placed in a row on each fide the midrib of the leaf, in large, yellow, round dots, ftanding on foot-ftalks, of a roundifh fhape, with the furface granulated from the feeds protuberating, furnifhed with a fhort elaftic fpring, and opening into two valves, *fig.* 2, 3, 4, 5, 6.

SEMINA plurima, ovata, aut fubreniformia flava, *fig.* 7, 8.

SEEDS numerous, oval or fomewhat kidney-fhaped, of a yellow colour, *fig.* 7, 8.

IN all thofe plants of the *Fern Tribe* which I have hitherto had an opportunity of examining, there appears to be much the fame mechanifm in their parts of fructification ; one of the moft ftriking and ufeful of which is the elaftic ring which furrounds the Capfules, by means of which they are forced open and the feeds difcharged. So neceffary a part one fhould not conceive would be wanting in any of thefe plants, nor will it, I believe, be found to be fo : yet many Botanifts, and thofe too of eminence, not only deny its exiftence, but make the want of it a character to diftinguifh this Genus. GLEDITCH gives us the following as part of the generic character of the *Polypodium* " *Capfulæ annulo deftitutæ.*" ADANSON alfo gives it the fame character, " *fans anneau.*" It will perhaps not be difficult to account for this miftake ; and at the fame time it will fhew us how injurious it is to fcience, for Authors to take things for granted without examining for themfelves. In TOURNEFORT's elegant figures of the Genera, the Capfules of the *Polypodium* are reprefented without any ring : on the truth of thefe figures it is highly probable that thofe Authors have relied ; for had they made ufe of their own eyes, affifted by a fmall magnifier, they could not have avoided feeing what MALPIGHI long before their time delineated, though rudely, and GLEICHEN fince more elegantly figured.

There is one circumftance attending this fpecies of *Polypodium*, which however does not run through the whole of this Genus, viz. the want of an Involucrum or Membrane; the little dots or affemblage of Capfules are not covered with any membrane ; or if there be a membrane, it is very early deciduous, and not vifible when the Capfules have arrived at a tolerable degree of maturity.

This fpecies of *Polypody* grows very common in woods and fhady lanes on the old ftumps of various trees ; it differs much in fize : fometimes it occurs on the Oak, in which cafe its virtue as a medicine has been more celebrated.

Its effects when taken inwardly are flightly purgative : it has been recommended in various diforders of the Vifcera, in the Cachexy, fwelling of the Spleen, Jaundice, obftructions of the Mefenteric Glands, Hypochondriac Difeafe, Cough, Afthma, &c. but it has generally been given with fome other medicines.

In the prefent practice it is but little regarded.

Bryum Scoparium.

BRYUM SCOPARIUM. BROOM BRYUM.

BRYUM *Linnæi Gen. Pl.* CRYPTOGAMIA MUSCI.

    Raii Syn. Gen. 3: MUSCI.

BRYUM *scoparium*, Antheris erectiusculis, pedunculis aggregatis, foliis secundis recurvatis, caule declinato.

    Linnæi Syst. Vegetab. p 797:

HYPNUM foliis falcatis, heteromallis; vaginis multifloris. *Haller hist. n.* 1777.

HYPNUM *scoparium. Scopoli Fl. Carn. p.* 334. DIAGN. Florescentia Hyemalis. Folia arcuata, secunda,

    tenuia. Setæ sæpe plures.

BRYUM *scoparium*: surculo declinato, ramoso, foliis secundis, recurvatis, primordialibus plumulosis. *Necker.*

    method. musc. p. 224.

HYPNUM *scoparium. Weis. Cryptogam. p.* 213.

BRYUM reclinatum, foliis falcatis, scoparum effigie: *The sickle-leaf'd bending Beasom Bryum. Dillen. musc. p.*

    357. *tab.* 46. *fig.* 16

BRYUM erectis capitulis angustifolium, caule reclinato. *Cat. Giss.* 221. *Raii Syn.* 95. *Hudson Fl. Angl. p.* 406.

CAULES unciales aut biunciales et ultra, suberecti, ramosi, in denso cæspite collecti, sordide rufi, infra multo tomento fusco obsiti.

STALKS from one to two inches high and more, nearly upright, branched, growing thickly together, of a dirty red colour, and covered at bottom with a dark brown wooly substance.

FOLIA caulem inæqualiter circumstant, hinc in quibusdam locis nudiusculus relinquitur, in aliis foliis crebrioribus vestitur, præcipue ad apicem, longa, linearia, acuminata, canaliculata, *fig.* 1, recurvata, secunda.

LEAVES : the leaves cover the stalk unequally, hence in some places it is left rather naked, in others more thickly covered with leaves, particularly towards the top, are long, linear, pointed, grooved, *fig.* 1, bent back, and turning all one way.

PEDUNCULI unciales aut biunciales, ad basin rubicundi, erecti, ex uno latere caulium plerumque oriuntur, aliquando vero ex apice, ut plurimum solitarii, subinde vero duo ex eodem perichætio proveniunt, basi bulbillo cylindraceo terminati, *fig.* 7, qui foliis pluribus latiusculis, pilo terminatis, acu facile separabilibus includitur, *fig.* 8, 9.

FOOT-STALKS an inch or two inches high, towards the bottom reddish, upright, arising generally from the side of the stalks, but sometimes from the top, most commonly single, but now and then two proceed from the same perichætium, furnished at bottom with a cylindrical bulb, *fig.* 7, which is inclosed by many broadish leaves, terminating in a hair, and easily separated by a needle, *fig.* 8, 9.

CAPSULÆ oblongæ et fere cylindraceæ, nunc erectæ, nunc paululum incurvatæ, *fig.* 3; OPERCULUM rostratum, tenue, longitudine capsulæ et concolor, *fig.* 4; ORA ciliata sive denticulata, *fig.* 5; CALYPTRA straminea, longitudine Capsulæ, postquam medio disrumpitur, basi suo capsulam arcte cingit, *fig.* 2; POLLEN viride, *fig.* 6.

CAPSULES oblong and almost cylindrical, sometimes upright, sometimes a little incurvated, *fig.* 3; the OPERCULUM the length of the Capsule. and of the same colour, terminating in a long slender point, *fig.* 4; the MOUTH ciliated or furnished with little teeth, *fig.* 5; the CALYPTRA straw-coloured, the length of the Capsule, after bursting in the middle closely embracing the Capsule by its base, *fig.* 2; the POLLEN green, *fig.* 6.

DILLENIUS very justly remarks, that this Moss seems to partake of the nature of both *Bryum* and *Hypnum*, but in his opinion, it comes nearest to the *Bryum*, and of the same sentiment appear to be LINNÆUS and NECKER, while HALLER, SCOPOLI, and WEIS, rank it among the *Hypnums*, and this they are led to, chiefly from the Peduncles being furnished at bottom with a kind of *Perichætium*; but DILLENIUS very properly observes, that although the peduncle is surrounded at bottom by many *squamæ* or *folioli*, yet these are not similar to what occur in the generality of *Hypnums*, as they may with the point of a pin be easily separated from one another, and then the base of the peduncle appears to be furnished with a *bulbillus* as in most of the *Bryums*: this circumstance added to its general habit, appears fully to justify this most excellent Botanist in placing it with the *Bryums*, from whence it ought not to have been separated without more weighty reasons than have been advanced.

This Moss distinguishes itself from most others by its beautifull and lively verdure; when young it puts on a very different appearance from what it has when farther advanced, being much shorter and its leaves upright; and DILLENIUS has sometimes remarked in this species *Stellulæ femineæ.*

It grows in very large Clumps or Patches forming a soft and delightfull Carpet, on the banks which surround woods, at the bottom of trees, and on heaths.

It is found on some parts of Hampstead heath producing its fructifications in February and March.

Bryum Undulatum.

Bryum undulatum. Curled Bryum.

BRYUM *Linnæi. Gen. Pl.* Cryptogamia Musci.

   Raii Syn. Gen. 3. Musci.

BRYUM *(undulatum)* antheris erectiufculis, pedunculis fubfolitariis, foliis lanceolatis carinatis undulatis patentibus ferratis. *Linn. Syft. Vegetab. p.* 797.

BRYUM foliis lanceolatis ferratis, capfulis cylindricis inclinatis ariftatis. *Haller. hift. tom.* 1. 1823.

BRYUM *phyllitidifolium:* furculo fimplici, foliis undato-ferrulatis, primordialibus plumulofis. *Neckeri method.
mufcor. p.* 203. cur nomen triviale a Cl. Neckero mutaretur non video, cum analogia unde nomen ejus fumitur obfcura fit, obfervante Cl. Scopoli.

BRYUM Phyllitidis folio rugofo acuto, capfulis incurvis. *Dillen mufc.* 360. *tab.* 46. *fig.* 18.

BRYUM undulatum. *Scopoli. Fl. Carniol. n.* 1301. *Raii Syn. p.* 95. 16. *Hudfon Fl. Angl.* 406. *Weis
Cryptogam.* 196. *Oeder Fl. Dan. tab.* 497. noftris duplo faltem minor, eum operculo nimis recto
et acuto.

SURCULI unciales, aut biunciales, plerumque fimplices, erecti, foliofi.

FOLIA lanceolata, undulata, carinata, *ferrato-aculeata,*
patentia, arefnctione involuta, *fig.* 1.

PEDUNCULI fimplices, (duo ex eodem furculononnun-
quam proveniunt) furculis plerumque longiores, erecti, rubri, *fig.* 2.

CAPSULA five Anthera cylindracea, incurvata, lente
vifa fubftriata, primum viridis, dein ex livido-
fufca, demum rufa, *fig.* 3. Bafis Operculi he-
mifphærica, rubra, apex pallida, fetacea, obtufiufcula, *fig.* 5, Capfulæ Ora ciliata Ciliis
inflexis, *fig.* 7; Annulus ruber, *fig.* 6; Pollen feu Semen viride, *fig.* 8.

CALYPTRA pallide fufca, acuminata, primum erecta,
flexurâ capfulæ difrumpitur, et recta manet, ba-
fique fuâ a Capfula fecedit, *fig.* 4.

STALKS from one to two inches high, generally fimple, upright and leafy.

LEAVES lanceolate, waved, keel-fhaped, *minutely and
fharply ferrated at the edges,* fpreading, when
dry curling in, *fig.* 1.

FOOT-STALKS of the frudification fimple, (fometimes two proceed from the fame ftalk) generally longer than the ftalks, upright, and of
a reddifh colour, *fig.* 2.

CAPSULE or Anthera cylindrical, incurvated, if
magnified appearing fomewhat ftriated; firft
green, then livid-brown, aud laftly of a reddifh
brown colour ; *fig.* 3, the bottom of the Operculum hemifpherical and red, the top paler,
very flender and rather blunt; *fig.* 5, the
Mouth of the Capfule furnifhed with Ciliæ
which bend inward, *fig.* 7 ; the Annulus or
Ring red, *fig.* 6 ; the Pollen or Seed green,
fig. 8.

CALYPTRA of a pale brown colour, and terminating
in a long point, firft upright, afterwards by the
bending of the Capfule it becomes burft at bottom, and remains ftrait, with its bafe at fome
little diftance from the Capfule.

This fpecies of Bryum is one of the largeft we have in this Country, it produces its frudification from
November to *February* and may be found in moft of the woods near Town, as well as on the heaths, but
more particularly in *Charlton Wood,* where it abounds.

As all its parts of frudification are large and diftinct, the botanic Student who would inveftigate this difficult
clafs of plants, cannot with this view, felect any mofs more proper for his purpofe.

Bryum hornum.

Bryum Hornum. Swans-neck Bryum.

MNIUM *Linnæi Gen. Pl.* Cryptogamia Musci. Masculus flos pedunculatus. Femineus flos in distincto

sæpius individuo.

Raii Synopsis Gen. 3. Musci.

MNIUM *hornum* antheris pendulis, pedunculo curvato, surculo simplici, foliolis margine scabris. *Linnæi Syst.*

Vegetab. 796.

BRYUM *hornum* surculo capitulifero ramosiusculo : stellifero simplici, primordialibus plumulosis. *Necker. Method.*

Musc. p. 215.

MNIUM foliis lanceolatis, imbricatis, capsulis pendulis cylindricis obtusis. *Haller. hist. helv.* 3. *p.* 54.

MNIUM *hornum* serratifolium. *Weis Cryptogam.* 149.

BRYUM antheris oblongis nutantibus, pedunculo curvato, foliolis ovatis, margine scabris. *Hudson. Fl.*

Angl. p. 415.

BRYUM stellare hornum sylvarum, Capsulis magnis nutantibus. *Dillen. musc.* 402.

BRYUM nitidum capitulis majoribus reflexis, calyptra imum vergente, pediculis oblongis e cauliculis novis

egredientibus. *Raii Syn. p.* 102. 51.

CAULES unciales aut biunciales, radiculis ferrugineis, valde tomentosis instructi, erecti, plerumque ramosi, pedunculiferi et stelliferi, ad basin rubicundi, Stellulæ et Pedunculi, nunc seorsim, nunc ex eadem radice proveniunt, unusque aut plures Surculi e basi caulis semper fere nascuntur.

FOLIA saturate viridia, ovato-lanceolata, suberecta, pellucida, ad lentem *minute serrata, fig.* 1 ; nervo medio distincto et in mucronem brevem educto, in surculis fœmineis dictis apice stellatim expansa, et paulo latiora, in junioribus angustiora et cauli magis adpressa.

PEDUNCULI terminales, biunciales, rubræ, versus apicem ut recte observavit Dillenius instar *colli olorini* incurvati.

CAPSULÆ *oblongæ*, tumidæ, virides, nutantes, lente aucta, *fig.* 7 ; per longitudinem secta ut Receptaculum conspiciatur, *fig.* 9 ; Calyptra longa, acuminata, caduca, *fig.* 6 ; Operculum breve, flavescens, *fig.* 8 ; Ora ciliata.

‡ This species comes hear to the largest size.

STALKS from one to two inches in height, furnished with roots which are of a ferruginous colour, and covered with a kind of wooly substance, upright and generally branched, reddish at bottom, producing both Pedunculi and Stellulæ, which proceed sometimes from the same, sometimes from different roots, and one or several Surculi usually spring from the bottom of the stalk.

LEAVES of a deep green colour, of an oval pointed shape, nearly upright, pellucid, when viewed with a glass *finely serrated at the edges, fig.* 1 ; the midrib distinct, and terminating in a short point, on the tops of those stalk, which are considered as female, they are expanded like a little star and somewhat broader, in the young shoots they are narrower and pressed closer to the stalk.

PEDUNCLES springing from the summit of the stalks, about two inches in height, bent near the top like a *Swans Neck* as Dillenius has properly observed.

CAPSULES *oblong*, tumid, of a green colour and drooping, magnified, *fig.* 7 ; cut longitudinally through the middle that the Receptaculum may be seen, *fig.* 9 ; the Calyptra long, pointed, and soon falling off, *fig.* 6 ; the Operculum short, of a yellowish colour, *fig.* 8 ; the Mouth of the Capsule ciliated.

On examining with a Microscope the tops of those Stalks which are called *Stellulæ femineæ, fig.* 2, and which are considered by many as the female parts of the fructification in this Moss, there appeared in the center of the *Stellula*, a great number of small upright bodies, or Corpuscles, of two kinds, *fig.* 3, the one white, pellucid, and jointed ; the other of a greener colour, shorter, and of an oblong oval shape, vid. *fig.* 4, 5. They do not appear to me to have any thing in their Structure, in the least similar to any of the parts of fructification in plants, what their real structure and uses are, may perhaps be discovered by future observations.

This species occurs not unfrequently on moist banks in Woods, as in *Charlton* Wood, and the Woods about *Hampstead*, producing its Fructifications in February and March.

As the *Capitula pulverulenta* of Dillenius, or *Sphærophylli* as they are called by Necker, are entirely wanting in this Moss, and as the existence of those singular little heads seems very obviously to distinguish the Genus *Mnium*, I have chosen rather to arrange it with Dillenius and Hudson among the *Bryums*, than with Linnæus among the *Mniums* ; for if we make *Mniums* of all the Mosses which have *Stellulæ*, we shall involve ourselves in considerable difficulties : many of those *Stellulæ* are indeed very obvious, as in the present one, but in others they are very obscure, so that it is difficult to say whether they exist in them or not ; but if they were obviously to be distinguished, there is not the least likeness between a *Stellula* and *Sphærophyllum*, why then unite in one Genus plants which have such very different appearances ? Would it not be better to consider the Mosses which produce *Sphærophylli* or *little balls* as *Mniums*, according to Dillenius, and divide the *Bryums*, if necessary, into two families, viz. such as have obvious *Stellulæ*, and such as have *none* ?

The name of *rough Bryum*, which Mr. Hudson seems to have given to this Moss for brevity's sake, conveys an idea with which this *Bryum* does not seem perfectly to correspond, it having no *roughness* except at the edges of the leaves, which are minutely serrated : I have therefore adopted Dillenius's name of *Swans Neck Bryum*, as being justifiable from the singular shape of the Peduncles, and being more likely to be remembered from its striking analogy.

Hypnum Proliferum. Proliferous Hypnum.

HYPNUM *Linnæi Gen. Pl.* Cryptogamia Musci.

    Raii Syn. Gen. 3. Musci.

HYPNUM *proliferum* furculis proliferis, plano-pinnatis, pedunculis aggregatis. *Linnæi Syst. Vegetab. p.* 860.

HYPNUM ramis teretibus pinnatis, pinnulis pinnatis, foliis adpreſſis. *Haller. hist.* 3. *p.* 33.

HYPNUM filicinum, Tamariſci foliis minoribus, non ſplendentibus. *Dillen. p.* 276. *icon.* 35. *fig.* 14.

HYPNUM repens filicinum minus, luteo virens. *Catal Giſſ.* 217. *Raii Synop. p.* 86. *n.* 36. *Hudſon, Fl. Angl. p.* 422. *Weis Cryptogam. p.* 230.

CAULES palmares ad dodrantales, repentes, hinc inde radiculas fuſcas exferentes, ſæpe vero adeo intricate connexi ut humi ſerpere nequeant, foliis ovato-acuminatis, carinatis, mucronatis, ſparſe tectis, *fig.* 1. horum foliolorum ſuperficies, microſcopio valde aucta granuloſa apparet, *fig.* 2.

RAMI pulchre pinnati, deflexi, vireſcentes, ad luteum colorem plus minuſve accedentes pro ratione ſitus aut anni temporis, *omni ſplendore deſtituti, rachis concolor*, ad extremitatem plerumque incraſſatus. Ramuli et Pinnulæ foliolis exiliſſimis, confertis, nudo oculo vix conſpicuis imbricatim tecti; e diſco rami, aut frondis, novus caulis aut furculus plerumque exſurgit, unde plantula mire extenditur ac propagatur, et hinc *Prolifer* vocatur.

PEDUNCULI ſeſquiunciales, rubri, plerumque quatuor aut quinque, aliquando plures e caule aggregatim aſſurgunt, et in quibuſdam caulibus, Perichætia plura aut potius eorum rudimenta occurrunt, e quibus Pedunculi ſequente anno probabiliter naſcuntur. Perichætium *fig.* 3. aut baſis pedunculi, ovatum, foliolis tenuibus pilo longo flexuoſo terminatis veſtitum. Capſulæ five Antheræ *fig.* 4. quæ Semen aut Pollinem continent, incurvatæ, ex fuſco aurantiacæ. Operculum *fig.* 6. (quod collo capſulæ inſigitur, et ſenſim maturaſcente decidit) breve, et acuminatum. Orificium Capſulæ duplici ſerie Ciliarum inſtruitur *fig.* 8. 9. Ciliæ *exteriores* fig. 8. aurantiacæ, divergentes, apicibus aliquando paululum inflexis, et cum aridæ ſint fragiles; *interiores* fig. 9. convergentes, membrana reticulata connexæ, ad quam videndam microſcopio opus eſt. Pollen five Semen viride. Calyptra *fig.* 5. quâ anthera cum ſuo Operculo partim tegitur et quæ primum decidit albida eſt.

STALKS from three to nine inches in length, creeping on the ground, and here and there ſending forth ſmall brown fibres, but very often ſo intricately connected together as to be hindered from creeping, thinly covered with leaves of an oval pointed ſhape, having a ſtrong midrib, which runs out to a fine point *fig.* 1. when greatly magnified the ſurface of theſe leaves exhibits a granulated appearance *fig.* 2.

BRANCHES beautifully pinnated, and bending downward, of a green colour, more or leſs inclined to yellow, according to its place of growth, and the ſeaſon of the year, *without any gloſs*; the *midrib of the ſame colour with the leaves* and generally thicker at its extremity; the ſmall leaves, laying one over another, and ſcarce diſcernible to the naked eye. From the middle of the branch or Frons moſt commonly ariſes a new ſtalk, or furculus, by which means this plant is ſingularly extended and propagated, and from this circumſtance it acquires the name of *Proliferous.*

PEDUNCLES about an inch and a half in length, of a bright red colour, generally about four or five, ſometimes more, ſpring from the ſtalk nearly together, in ſome of the ſtalks there is the appearance of ſeveral *Perichætia* without peduncles, which probably ariſe from them the next year. The Perichætium *fig.* 3. which is the baſe of the peduncle, is of an oval ſhape, and covered with ſmall leaves which terminate in a long flexible point. The Capsules or Antheræ containing the pollen or ſeed *fig.* 4. are incurvated, and of a brown orange colour. The Operculum *fig.* 6. (which fits on to the top of the Capſule, and when the ſeed contained within it, is ripe, falls off) is ſhort, and pointed; the mouth of the Capſule has two rows of Ciliæ *fig.* 8. 9; the exterior row *fig.* 8. orange coloured, and diverging, the tops of them ſometimes bending a little inward, and brittle when dry, the interior row *fig.* 9. converging, of a membranous texture, and when very much magnified, appearing reticulated. The Pollen or Seed contained within the Capſules is green. The Calyptra *fig.* 5. which partly covers the anthera and operculum, and firſts drops off is of a white colour.

There is ſcarce a Wood in the environs of this City, on the borders of which this elegant ſpecies of Moſs doth not occur.

It produceth its fructifications from December to February; in this ſtate however it is but ſeldom met with, yet may be found by diligent ſearching. Linnæus in one of his journies through Sweden, obſerved this Moſs growing in the thickeſt Woods, obſcured with perpetual ſhade, and where all other plants periſhed.

Moſt of the writers who have made this claſs of plants more particularly the object of their enquiries, have generally made two diſtinct Genera of the Hypnum and Bryum, yet ſo great is the affinity betwixt them, and ſo much do they run into one another, that what ſome of theſe Authors call a Bryum, others denominate a Hypnum; indeed this diviſion ſeems adopted more to facilitate the inveſtigation of the plants of this numerous family, than from any real natural diviſion which takes place between them. The difference between ſome of the Hypnums and ſome of the Bryums is obvious to almoſt every one, but to aſcertain the limits where the one begins and the other terminates, ſeems a taſk too difficult for the moſt accurate Botaniſt.

The principal Characteriſtics of a Bryum according to Linnæus, are, that the peduncle which ſuſtains the Anthera or Capſule, grows out of the top of the furculus or ſtalk, and is furniſhed at its baſe with a little naked tubercle or bulb; in the Hypnum on the contrary, the peduncle grows out of the ſide of the ſtalk and the tubercle at its baſe is covered with leaves and called a Perichætium.

www.ingramcontent.com/pod-product-compliance
Lightning Source LLC
Chambersburg PA
CBHW021514210326
41599CB00012B/1248